Solar Electricity Handbook

2021 Edition

A simple, practical guide to solar energy: how to design and install photovoltaic solar electric systems

www.SolarElectricityHandbook.com

Michael Boxwell

Greenstream Publishing
Unit 51 Imex Business Park
Birmingham
B11 2AL
United Kingdom

www.greenstreampublishing.com

Published by Greenstream Publishing, 2021.

Copyright © 2009–2021 Michael Boxwell.
All rights reserved.

Paperback Printed Edition: ISBN 978-1-907670-75-6

Also available in hardback and ebook formats.

Fourteenth Edition, published January 2021

Michael Boxwell asserts the moral right to be identified as the author of this work.

A catalogue record for this book is available from the British Library.

While we have tried to ensure the accuracy of the contents of this book, the author or publishers cannot be held responsible for any errors or omissions found therein.

All rights reserved. No part of this publication may be reproduced, stored in a retrieval system, or transmitted, in any form or by any means, electronic, mechanical, photocopying, recording or otherwise, without the prior permission of the publishers.

TABLE OF CONTENTS

Introducing Solar Energy 1
Making and saving money with solar 17
A Brief Introduction to Electricity 22
The Four solar pv Configurations 28
Components of a Solar Electric System 36
The Design Process 44
Scoping the Project 47
Calculating Solar Energy 56
Surveying Your Site 74
Understanding the Components 98
Components for Grid-Tie systems 116
Components for Stand-Alone Systems 128
Planning, regulations and approvals 171
Detailed Design 174
Installation 201
Troubleshooting 214
Maintaining Your System 222
Internet Support 224
Appendix A – Typical Power Requirements 226
Appendix B – Living Off-Grid 229
A Final Word 236

INTRODUCING SOLAR ENERGY

Ninety-three million miles from Earth, our sun is 1,300,000 times the size of our planet. With a diameter of 865,000 miles, a surface temperature of 5,600°C (over 10,000°F) and a core temperature of 15,000,000°C, it is a huge mass of constant nuclear activity.

Both directly and indirectly, our sun provides all the power we need to exist and supports all life on earth. The sun drives our climate and our weather. It's a huge energy source. Without it, our world would be a frozen wasteland of ice-covered rock.

Finding ways to harness the sun's energy and using it to power electrical equipment is a terrific idea. There are no ongoing electricity bills, no reliance on a power socket, simply a free and everlasting source of energy that does not harm the planet. Of course, the reality is a little different from that and solar energy will not work for everyone. Yet generating electricity from sunlight alone is a powerful resource, with applications and benefits throughout the world.

So how does it work? For what is it suitable? What are the limitations? How much does it cost? How do you install it? This book answers all these questions and shows you how to use the power of the sun to generate electricity yourself.

Along the way, I will also expose a few myths about some of the wilder claims made about solar energy and I will show you where solar power may only be part of the solution.

I will keep the descriptions as straightforward as possible. There is some simple mathematics and science involved. This is essential to allow you to plan a solar electric installation successfully. However, none of it is complicated and there are plenty of short-cuts and online calculators at *www.SolarElectricityHandbook.com* available to keep things simple.

The book includes example projects to show how you can use solar electricity. Some of these are very straightforward, such as providing electrical light for a shed or garage, for example, or fitting a solar panel to the roof of a caravan or boat. Others are more complicated, such as installing photovoltaic solar panels to a house.

I also show some rather more unusual examples, such as the possibilities for solar electric motorbikes and cars. These are examples of what can be achieved using solar power alone, along with a little ingenuity and determination.

I have used one main example throughout the book: providing solar-generated electricity for a holiday home which does not have access to an electricity supply from the grid. I have created this example to show the issues and pitfalls that you may encounter along the way, based on real life issues and practical experience.

A website accompanies this book. It has lots of useful information, along with a suite of online solar energy calculators that will simplify the cost analysis and design processes. The website is at *www.SolarElectricityHandbook.com*.

WHO THIS BOOK IS AIMED AT

If you simply want to gain an understanding about how solar electricity works, then this handbook will provide you with everything you need to know.

If you are planning to install your own stand-alone solar power system, this handbook is a comprehensive source of information that will help you understand solar and guide you in the design and installation of your own solar electric system.

Solar has a big application for integrating into electrical products such as mobile phones, laptop computers and portable radios. Even light electric cars can use solar energy to provide some of their power requirements, depending on the application. If you are a designer, looking to see how you can integrate solar into your product, this book will give you a grounding in the technology that you will need to get you started.

If you are specifically looking to install a grid-tie system, i.e. a solar energy system that will feed electricity back into your local power grid, this book will provide you with a good foundation and will allow you to carry out the design of your system. You will still need to check the local planning laws and any other local legislation surrounding the installation of solar energy systems, and you will have to understand the building of electrical systems. In some countries, you need to be certified to carry out the physical installation of a grid-tie system. You'll also need a good understanding of electrical systems, with the relevant qualifications to carry out the work.

If you are planning to install larger, commercial-size systems, or hoping to install grid-tie solar systems professionally, then this book will serve as a good introduction, but you will need to grow your knowledge further. This book gives you the foundations you need to get started, but there are special skills required when designing and implementing larger scale solar systems that go far beyond what is required for smaller systems and are beyond the scope of this book.

If you are a hobbiest planning your own solar installation, it will help if you have some DIY skills. Whilst I include a chapter that explains the basics of electricity, a familiarity with wiring is also of benefit for smaller projects and you will require a thorough understanding of electrical systems if you are planning a larger project such as powering a house with solar.

THE RAPIDLY CHANGING WORLD OF SOLAR ENERGY

I implemented my first solar energy system in 1997. Back then, the largest panels I could buy were 100Wp panels. Each panel was around 9% efficient. The cost for the ten panels I needed to create my 1kWp array was almost £10,000 (around $13,000).

Today, solar panels can be up to 24% efficient. They have a far higher capacity, a longer life expectancy and they are far cheaper. If I were to recreate my first 1kWp array today using larger, modern solar panels, the panel price would be around £450 ($590). Ignoring 23 years of inflation, that is a 96% drop in the ticket price for a solar panel!

Of course, there is more to the installation cost than just the solar panels alone. Cabling, inverters, batteries, mounting brackets and time all cost money, and most of those prices have increased in the past twenty years, but the fact remains that solar has been transformed from an obscure niche product to mainstream energy generation because of incredible leaps in the technology that has made solar more efficient and more affordable.

I wrote the first edition of this book early in 2009. This 2021 issue is the fourteenth edition. Most editions have included significant rewrites to keep up with the rapid pace of change, and this edition is no exception.

The rapid improvement in the technology and the freefall in costs since early 2009 have transformed the industry. Systems that were completely unaffordable or impractical five or six ago are now cost-effective and achievable.

Solar panels available today are smaller, more robust and better value for money than ever before. Battery storage has become more reliable and far cheaper. For many more applications, solar is now the most cost-effective way to generate electricity.

Government incentives to promote renewable energy sources have also changed why people install solar. In countries like Germany, the United Kingdom, Spain and in the southern states of the United States, residential solar is becoming a common sight. Solar is making a big impact in the way that countries generate electricity. In several countries across Europe, solar panels are producing as much as 50% of the nation's electricity during peak production periods in summer.

Over the coming years, all the signs are that the technology and the industry will continue to evolve at a similar pace. In the 2012 edition I claimed that solar will be the cheapest form of electricity generator by 2015, undercutting traditionally low-cost electricity generators such as coal-fired power stations. I was wrong. We reached that cross-over point in many parts of the world by the middle of 2013. Huge solar farms are becoming a common sight, not just in wealthy parts of the world with hot, sunny climates such as in California and the southern parts of Europe, but in Canada and northern Europe where weather is less predictable and in India and China, where solar was an unaffordable luxury only two or three years ago.

Between 2009 and early 2014, prices of solar panels were in freefall, dropping by as much as one-third each year. Since then solar panel prices have continued to fall by around 10% per year. Consequently, the cost of a solar panel today is $1/_8$th of its price eight or nine years ago.

As prices have fallen and efficiencies improved, we have seen solar energy incorporated into more everyday objects such as laptop computers, mobile phones, backpacks and clothing. Meanwhile, solar energy is causing a revolution for large areas of Asia and Africa, where entire communities are now gaining access to electricity for the first time.

As an easy-to-use and low-carbon energy generator, solar is without equal. Its potential for changing the way we think about energy in the future is huge. For families and businesses in rural African and Asian villages, it is creating a revolution and saving lives.

SOLAR ELECTRICITY AND SOLAR HEATING

Solar electricity is produced from sunlight shining on photovoltaic solar panels. This is different to solar hot water or solar heating systems, where the power of the sun is used to heat water or air.

Solar heating systems are beyond the remit of this book. That said, there is some useful information on surveying and positioning your solar panels in this book that is relevant to both solar photovoltaics and solar heating systems.

If you are planning to use solar power to generate heat, solar heating systems are far more efficient than solar electricity, requiring far smaller panels to generate the same amount of energy.

Solar electricity is often referred to as photovoltaic solar, or PV solar. This describes the way that electricity is generated in a solar panel. For the purposes of this book, whenever I refer to *solar panels* I am talking about photovoltaic solar panels for generating electricity, and not solar heating systems.

THE SOURCE OF SOLAR POWER

Deep in the centre of the sun, intense nuclear activity generates huge amounts of radiation. In turn, this radiation generates light energy called photons. These photons have no physical mass of their own but carry huge amounts of energy and momentum.

Different photons carry different wavelengths of light. Some photons will carry non-visible light (*infra-red* and *ultra-violet*), whilst others will carry visible light (referred to as *white light*).

Over time, these photons push out from the centre of the sun. It can take one million years for a photon to push out to the surface from the core. Once they reach the sun's surface, these photons rush through space at a speed of 670 million miles per hour. They reach earth in around eight minutes.

On their journey from the sun to earth, photons can collide with and be deflected by other particles and are destroyed on contact with anything that can absorb radiation, generating heat. Your body absorbs photons from the sun. That is why you feel warm on a sunny day.

Our atmosphere absorbs many of these photons before they reach the surface of the earth. That is one of the two reasons that the sun feels so much hotter in the middle of the day. The sun is overhead, and the photons have to travel through a thinner layer of atmosphere to reach us, compared to the end of the day when the sun is setting, and the photons have to travel through a much thicker layer of atmosphere.

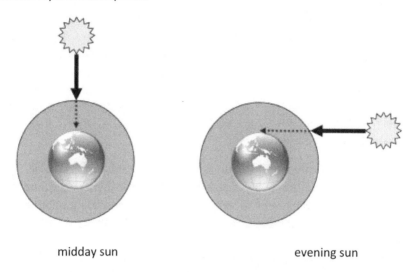

midday sun   evening sun

This is also one of the two reasons why a sunny day in winter is so much colder than a sunny day in summer. In winter, when your location on the earth is tilted away from the sun, the photons must travel through a much thicker layer of atmosphere to reach us. The other reason that the sun is hotter during the middle of the day than at the end is because the intensity of photons is much higher at midday. When the sun is low in the sky, these photons are spread over a greater distance simply by the angle of your location on earth relative to the sun.

THE PRINCIPLES OF SOLAR ELECTRICITY

A solar panel generates electricity using the *photovoltaic effect*, a phenomenon discovered in 1839 when Edmond Becquerel, a French physicist, observed that certain materials produced an electric current when exposed to light.

Two layers of a semi-conducting material are combined to create this effect. One layer must have a depleted number of electrons. When exposed to sunlight, the layers of material absorb the photons. This excites the electrons, causing some of them to 'jump' from one layer to the other, generating an electrical charge.

The semi-conducting material used to build a solar cell is silicon, cut into very thin wafers. Some of these wafers are then 'doped' to contaminate them, thereby creating an electron imbalance in the wafers. The wafers are then aligned together to make a solar cell. Conductive metal strips attached to the cells take the electrical current.

When a photon hit the solar cell, it can do one of three things. It can be absorbed by the cell, reflected off the cell or pass straight through the cell. When a photon is absorbed by the silicon, electrical current is generated. The more photons (i.e. the greater intensity of light) that are absorbed by the solar cell, the greater the current generated.

Solar cells generate most of their electricity from direct sunlight. However, they also generate electricity on cloudy days and some systems can even generate very small amounts of electricity on bright moonlit nights.

Individual solar cells typically only generate tiny amounts of electrical energy. To make useful amounts of electricity, these cells are connected together to make a solar module, otherwise known as a solar panel or, to be more precise, a photovoltaic module.

UNDERSTANDING THE TERMINOLOGY

In this book, I use various terms such as 'solar electricity', 'solar energy' and 'solar power'. Here is what I mean when I am talking about these terms:

Solar power is a general term for generating power, whether heat or electricity, from the power of the sun.

Solar energy refers to the energy generated from solar power, whether electrical or as heat.

Solar electricity refers to generating electrical power using photovoltaic solar panels.

Solar heating refers to generating hot water or warm air using solar heating panels or ground-source heat pumps.

SETTING EXPECTATIONS FOR SOLAR ELECTRICITY

Solar power is a useful way of generating reasonable amounts of electricity, so long as there is a good amount of sunlight available and your location is relatively free from obstacles such as trees and other buildings that will shade the solar panel from the sun.

Solar can be used in various ways. For instance, you can use solar power as the only source for electricity, in which case you need to ensure that your solar panels provide enough energy to handle all your requirements, and you will need batteries to store the energy in. Alternatively, you can use solar power to supplement your electricity supply, such as installing solar on the roof of your home, using solar power during the day and your existing electricity supply for power each night.

When I first became involved in solar in the mid-1990s, most applications were either to add power generation for a specific appliance or application, such as a telecoms transmitter on the top of a hillside where electricity was not available, or to add electricity generation to a product, such as a pocket calculator or a torch. Installing solar onto a building was almost unthinkable because of the cost, and when solar was installed, it was usually to power specific circuits within the house, such as lighting, rather than incorporated into the entire household electrics.

Today, the main uses for solar are either on a building to supplement an existing electricity connection, or in huge solar farms, generating electricity to feed onto the main electricity grid, either directly or in combination with a battery bank.

Solar as your only source of electricity

If you are looking to use solar as your only source of electricity, you need to take a good hard look at your consumption and see how you can keep demand to a minimum. In a home environment, it is quite tricky to do this without making big changes. As consumers, it is very easy to underestimate how much electricity we use, and solar power can end up becoming very expensive if you simply want to match your typical energy consumption without looking to make savings.

Of course, it is possible to put in a cheap and simple solar energy system, which doesn't cost a fortune and can give you decades of reliable service. Some examples include:

- Installing a light or a power source somewhere where it is tricky to get a standard electricity supply, such as in the garden, shed or remote garage
- Creating a reliable and continuous power source where the standard electricity supply is unreliable because of regular power cuts

- Building a mobile power source that you can take with you, such as a power source for use whilst camping, working on outdoor DIY projects or working on a building site

The amount of energy you need to generate has a direct bearing on the size and cost of a solar electric system. The more electricity you need, the more difficult and more expensive your system will become.

If your requirements for solar electricity are to run a few lights, to run some relatively low-power electrical equipment such as a laptop computer, a small TV, a compact fridge and a few other small bits and pieces, then if you have a suitable location you can achieve what you want with solar.

On the other hand, if you want to run high-power equipment such as fan heaters, washing machines and power tools, you are likely to find that the costs will rapidly get out of control.

As I mentioned earlier, solar electricity is not well suited to generating heat. Heating rooms, cooking and heating water all take up significant amounts of energy. Using electricity to generate this heat is extremely inefficient. Instead of using solar electricity to generate heat, you should consider a solar hot water heating system, and heating and cooking with gas or solid fuels.

It is not normally possible to power the average family home purely on solar electricity without making any cuts in your current electricity consumption. Most houses do not have enough roof space for all the panels that would be required. Cost is also a factor. If you are planning to go entirely off-grid, it is usually a good idea to carefully evaluate your electricity usage and make savings where you can before you proceed.

Most households and businesses are very inefficient with their electrical usage. Spending some time first identifying where electricity is wasted and eliminating this waste is an absolute necessity if you want to implement solar electricity cost-effectively.

This is especially true if you live in cooler climates, such as Northern Europe or Canada, where the winter months produce much lower levels of solar energy. In the United Kingdom, for instance, the roof of the average-sized home is not large enough to hold all the solar panels that would be required to provide the electricity used by the average household in the depths of winter. In this instance, making energy savings is essential.

For other applications, a solar electric installation is much more cost-effective. For instance, no matter which country you live in, providing electricity for a small holiday home is well within the capabilities of a solar electric system, so long as heating and cooking are catered for using gas or solid fuels and the site is in a sunny position with little or no shade. In this scenario, a solar electric system may be more cost-effective than installing a conventional electricity supply if the house is off-grid and is not close to a grid electricity connection.

If your requirements are more modest, such as providing light for a lock-up garage or a stable for horses, for example, there are off-the-shelf packages to do this for a very reasonable cost. Around £20–£80 ($30–$130) will provide you with a lighting system for a shed or small garage, whilst £200 ($260) will provide you with a system big enough for lighting large stables or a workshop.

This is far cheaper than installing a conventional electricity supply into a building, which can be expensive even when a local supply is available just outside the door.

Low-cost solar panels are also ideal for charging up batteries in caravans and recreational vehicles or on boats, ensuring that the batteries get a trickle charge between trips and keeping the batteries in tip-top condition whilst the caravan or boat is not in use.

Solar as a supplemental power source

Most solar installations in Europe and North America today are roof-top installations on homes. In these systems, the solar panels generate electricity, which is used within the home during the day with any excess energy being fed into the grid and used by other homes nearby. At night, these homes then use electricity from the utility grid. These systems are called 'grid-tie' systems.

Grid-tied solar electric systems effectively create a micro power station that provides energy not just for yourself, but for your local community. Electricity can be used by other people as well as yourself.

There are a few variations on this theme. For example, you can now install a battery with a grid-tie system so that excess electricity being produced during the day can be stored in a battery for use overnight. This has the advantage of reducing your personal dependence on the grid, but still provides the assurance that you will not run out of energy.

When configuring solar as a supplemental power source, the size of your solar installation is far less critical. Instead, you can choose the size of system based on the amount of space available, or by the amount of budget available for your project.

Payback for these systems varies considerably based on where you live, how much sunlight you get and how much electricity you use during the day. If you are home-based, have air conditioning or electric based heating, or can charge an electric car during the day, a solar installation that should last many decades can pay for itself in as little as four to six years. For most people, payback takes a little longer, but a typical home solar installation will usually pay for itself in around seven to ten years.

WHY CHOOSE A SOLAR ELECTRIC SYSTEM?

There are many reasons to consider installing a solar electric system:

- Where there is no other source of electrical power available, or where the cost of installing conventional electrical power is too high.
- Where other sources of electrical power are not reliable. For example, when power cuts are an issue and a solar system can act as a cost-effective contingency.
- When a solar electric system is the most convenient and safest option. For example, installing low voltage solar lighting in a garden or providing courtesy lighting in a remote location.
- When you can become entirely self sufficient with your own electrical power.
- When you want a mobile power source – such as in a caravan, travel trailer, recreational vehicle or boat.
- When there is sufficient financial incentive through savings in electricity being purchased from your electricity supplier to justify the investment.
- Once installed, solar power provides virtually free power without damaging the environment.

COST-JUSTIFYING SOLAR

Calculating the true cost of installing a solar electric system depends on various factors:

- The power of the sun at your location at different times of the year.

- How much energy do you need to generate?
- How good is your site is for capturing sunlight?

Despite the huge drop in prices over the past ten years, solar electric systems can often have a higher capital cost than some other sources. For example, a small generator will be cheaper to buy than an equivalent solar package of panel, a controller, battery and inverter. However, when you consider the running costs, solar invariably wins out. To create a comparison with alternative power sources, you will sometimes need to calculate a payback of costs over time to justify the initial cost of a solar electric system.

On all but the simplest of installations, you will need to carry out a survey on your site and carry out some of the design work before you can ascertain the total cost of installing a photovoltaic system. Do not panic, this is not as frightening as it sounds. It is not difficult; I cover it in detail in later chapters. We can then use this figure to put together a cost-justification on your project to compare with the alternatives.

SOLAR POWER AND WIND POWER

Wind turbines can be a good alternative to solar power, but probably achieve their best when implemented *together* with a solar system. A small wind turbine can generate electricity in a breeze, both day and night, and typically generates more energy in the winter months. Consequently, wind turbines and solar panels can compliment each other very well in the right application.

Small wind turbines have some disadvantages. Firstly, they are very site-specific, requiring higher than average wind speeds and minimal turbulence. They must be mounted so that the blades are at least 10m (32 feet) higher than their surroundings and away from tall trees. If you live on a windswept farm or close to the coast, a wind turbine can work well. If you live in a built-up area or close to trees or main roads, you will find a wind turbine unsuitable for your needs.

Compared to the large wind turbines used by the power companies, small wind turbines are not particularly efficient. If you are planning to install a small wind turbine in combination with a solar electric system, a smaller wind turbine that generates a few watts of power at lower wind speeds is usually better than a large wind turbine that generates lots of power at high wind speeds.

GENERATORS

For off-grid power that can be relied upon throughout the year, many people combine solar with batteries, and use a generator to provide back-up power, particularly during the winter months. Whilst running a diesel or petrol generator may not be the most environmentally friendly form of power generation, when used to supplement solar, they can be both cost effective and provide a reliable source of power. Furthermore, as the generator set up can be optimized to recharge batteries as well as provide instantaneous power, the generator can be tuned to run at its peak performance settings, running for short periods of time and generating power as efficiently as possible.

FUEL CELLS

As an alternative to generators, fuel cells can be a good way to supplement solar energy, especially for solar electric projects that require additional power in the winter months, when solar energy is at a premium.

A fuel cell works like a generator. It uses a fuel mixture such as methanol, hydrogen or zinc to create electricity.

Unlike a generator, a fuel cell creates energy through chemical reactions rather than through burning fuel in a mechanical engine. These chemical reactions are far more carbon-efficient than a generator.

Fuel cells are extremely quiet, although rarely completely silent, and produce water as their only emission. This makes them suitable for indoor use with little or no ventilation.

Unfortunately, fuel cells are expensive to buy and run and are therefore still a niche application.

SOLAR ELECTRICITY AND THE ENVIRONMENT

Emissions and the environmental impact of power generation is high on the agenda of all nations. Since the COP21 World Climate Summit held in Paris during December 2015, countries are committed to hitting tough targets. And despite President Trump's more recent changes of policy, the world remains largely united in its commitment to improving our environment. To achieve these goals, there must be a big shift away from traditional power generation towards greener energy production.

The energy industry is responding. China is building and installing two grid-scale wind turbines per hour, the United Kingdom is shutting down its coal-fired power stations years ahead of schedule and many countries have committed to eradicating fossil fuels from their power mix over the next five years.

No power generation technology is entirely environmentally friendly. Hydro-electric power stations have an impact on water courses and impacts local wildlife. Wind turbines account for bird deaths every year. Building hydro, wind or solar equipment also has a carbon footprint that must be accounted for. Yet this environmental impact is a tiny fraction of the carbon footprint associated with more traditional power generation technologies.

Once installed, a solar electric system is a low-carbon electricity generator. The sunlight is free and the system maintenance is extremely low. There is a carbon footprint associated with the manufacture of solar panels, and in the past this footprint has been quite high, mainly due to the relatively small volumes of panels being manufactured and the chemicals required for the 'doping' of the silicon in the panels.

Thanks to improved manufacturing techniques and higher volumes, the carbon footprint of solar panels is now much lower. You can typically offset the carbon footprint of building the solar panels by the energy generated within one to two years, and some of the very latest amorphous thin-film solar panels can recoup their carbon footprint in as little as six months.

Therefore, a solar electric system that runs as a complete stand-alone system can reduce your carbon footprint, compared to taking the same power from the grid.

The same is true of grid-tie solar. In the past, the power companies have struggled to integrate renewable energy into the mix of power generation sources. This has meant that whilst the energy produced by solar panels was non-polluting, it did not necessarily mean that there was an equivalent drop in carbon production at a coal or gas-fired power station.

That is no longer the case. Power companies have become far better at predicting weather conditions in advance and tuning the general mix of power generators to take advantage of renewable energy sources. This has ensured a genuine carbon reduction in our energy mix.

Of course, the sunnier the climate, the bigger benefit a solar energy system has in reducing the carbon footprint. In a hot, sunny region, peak energy consumption tends to occur on sunny days as people try to keep cool with air conditioning. In this scenario, peak electricity demand occurs at the same time as peak energy production from a solar array, and a grid-tie solar system can be a perfect fit.

If you live in a cooler climate with less sunshine, peak energy demand often occurs in the evening, when solar energy production is dropping. This does not negate the carbon benefit from installing solar, but if you want to maximize the carbon benefit of a solar energy system, you should try to achieve the following:

- Use the power you generate for yourself
- Use solar energy for high load applications such as clothes washing
- Reduce your own power consumption from the grid during times of peak demand
- Store some of the excess solar energy production using batteries and use it in the evening

Environmental efficiency: comparing supply and demand

There is an online calculator to map your electricity usage over a period of a year and compare it with the amount of sunlight available. Designed specifically for grid-tie, this calculator shows how close a fit solar energy is in terms of supply and demand.

Whilst this online calculator is no substitute for a detailed electrical usage survey and research into the exact source of the electricity supplied to you at your location, it will give you a good indication of the likely environmental performance of a solar energy system.

To use this online calculator, collate information about your electricity usage for each month of the year. You will usually find this information on your electricity bill or by accessing your electricity account online. Then visit *www.SolarElectricityHandbook.com*, follow the links to the Grid-Tie Solar Calculator in the Online Calculators section.

IN CONCLUSION

- Solar electricity can be a great source of power where your power requirements are modest, there is no other source of electricity easily available and you have a good amount of sunshine.
- Solar electricity is not the same as solar heating.

- Solar panels absorb photons from sunlight to generate electricity. Direct sunlight generates the most electricity, but solar still generates power on dull days.
- Solar electricity will not generate enough electricity to power the average family home all of the time, unless major economies in the household power requirements are made.
- Larger solar electric systems have a comparatively high capital cost, but the ongoing maintenance costs are very low.
- Smaller solar electric system can be extremely cost-effective to buy and install, even when compared to a conventional electricity supply.
- It can be much cheaper using solar electricity at a remote building, rather than connecting it to a conventional grid electricity supply.
- Both stand-alone and grid-tie solar energy systems can have a big environmental benefit.

MAKING AND SAVING MONEY WITH SOLAR

Creating energy with solar is not only environmentally friendly, it can also be good for your bank account too. Whether you want to offset your electricity bill in your home or business, or if you want to avoid the high cost of connection to the electricity grid for a new building, solar can often save you money.

In addition, there are often subsidies, grants or other financial incentives available to make solar a more attractive purchase. In some cases, these incentives alone are sufficient to pay for your solar installation over a period of a few years.

THE EVER-RISING COST OF ENERGY

The world faces an energy crisis. Traditional forms of electricity production, using coal, oil or gas, may be comparatively cheap, but come at a big cost. The cost to the environment is huge, fossil fuel price fluctuations creates uncertainty and the security of our energy supplies is a constant fear, responsible for more wars around the world than most politicians would care to admit to.

Oil and gas prices fluctuate wildly. The price of a barrel of crude oil can fluctuate by $10 in a single day. Since 2010, a barrel of crude has been as low as $17 and as high as $156.

The volatility in the market has proven difficult to predict. At the beginning of 2016, oil prices were as little as $22 per barrel. Oil experts were predicting a glut of cheap oil for the rest of the decade. Industry analysts were predicting that oil would therefore remain the cheapest form of energy for many years to come and shares in solar companies dropped significantly.

Yet one year later, oil prices had more than doubled to $45, increased to $58 per barrel by the time I came to write the 2018 edition of this book, and $85 per barrel one year later.

During the early stage of the COVID-19 pandemic, oil prices collapsed as nation after nation went into lockdown. At one point, crude oil prices dropped below zero as producers were unable to find storage facilities for their product.

Right now, in early 2021, oil prices are trading at around $51, but have fluctuated between $18 and $75 in the past quarter alone. As ever, oil prices remain highly volatile and the recent COVID-19 virus outbreak is destabilizing prices more than ever. The changes in oil prices makes and breaks economies, with countries like Peru, Russia and Saudi Arabia on an economic knife-edge due to fluctuating oil prices. Inevitably it is you and I who pick up the bill.

Most countries recognize the problem of price instability and have been implementing renewable energy schemes to reduce our reliance on fossil fuels. Yet this too comes at a price. Investment into new power stations and new technologies for energy production is expensive. Many countries, including the United Kingdom, Australia and the United States, have aging power stations and have suffered from decades of under investment into new technologies. Whilst this is now changing, the costs for these changes are huge and are reflected in the price we are paying for our energy.

In the United Kingdom, average consumer gas and electricity prices today are around two-and-a-half times higher than they were ten years ago. Most energy experts predict above-inflation rises for the next decade or more as the country phases out old coal-fired power station and replaces them with new power sources. In the United States, energy prices for the home has almost doubled over the same period. New taxes for above average electricity consumption in states such as Alabama, Texas and California being introduced in 2021 are going to penalize a lot of homeowners this year. The outlook for the next ten years suggests prices could easily double again.

It is because of this constant increase in energy prices that solar can become cost effective, both for many home and small business owners and for commercial and industrial users, too. In both Europe and the US, many hotel and restaurant chains are now choosing to install solar on their roofs and there is a renewed interest in solar farms, with typical paybacks of around eight years in Northern Europe and five years in Southern Europe and the southern states in the USA.

OUR GREATER RELIANCE ON ELECTRICITY

In the drive to reduce our carbon footprint, we are moving away from gas and oil and towards electricity for more of our energy usage. Building heating is moving from gas boilers towards electric heat pumps; whilst cars, vans and buses are becoming electric. Yet with retail electricity costs significantly higher than gas, keeping the cost of bills down is becoming more difficult. Installing solar can make a significant difference, making a long-term and constant saving on energy bills.

INSTALLING ELECTRICITY TO A NEW BUILDING

If you are looking to install an electricity connection to a new building, installing your own solar power station can often be cheaper than installing the power lines. This is particularly true if you live in a rural location where power lines may not run close to the building. Even simple connections to a roadside property can easily cost several thousand, and if your property is suitable for solar power it can quite easily become more cost effective to go completely 'off grid' and create all your own power from solar.

Of course, there are limitations to this approach. You need to produce *all* the energy that you use, and you will need a watchful eye on your electricity usage to make sure you do not run out. You may need to supplement solar with other forms of power generation, such as a wind turbine or a small generator for emergency use. Yet this can be a practical option for many locations where a conventional electricity connection is otherwise unavailable or unaffordable.

FINANCIAL INCENTIVES

Many countries brought in subsidy packages, grants and other financial incentives in order to encourage the uptake of solar. As prices have fallen, however, many of these subsidies have now been scaled back or withdrawn completely.

It is possible to generate income from solar through the sale of electricity through an export tariff. If you are putting in a large-scale solar farm, this income can pay for the full installation over a five- to ten-year period, possibly even shorter if combined with battery storage. For small scale solar, however, the income from selling surplus electricity back to the grid is unlikely to be a significant amount. In the UK, for example, a 4kWp solar installation is likely to generate less than £160 a year through the sale of electricity back to the grid.

Grant Schemes, Low Interest Loans and Tax Rebates

Grant schemes, low interest loans and tax rebates are paid for directly out of government funds. Consequently, they can be unpopular with some politicians and the wider electorate, who often resent tax-payers money being used as a green energy subsidy.

Grants for installing solar are now becoming much rarer, although a few schemes are still available around the world for specific applications:

- The United States Department of Agriculture and Rural Development has guaranteed loans and grants available for the installation of solar for agricultural and rural small businesses.
- The United Kingdom Rural Development Programme introduced a new programme in 2019 to help encourage the uptake of solar energy systems within the agriculture community, and this has been particularly popular in remote areas where suitable electricity supplies have not been available. This fund consists of a grant of up to £20,000 to test the viability of a renewable energy system, followed by a low-interest, unsecured loan to fund up to 50% of the installation costs for a wind, solar or hydro-power project.
- Also, in the United Kingdom, the Green Deal loan and Green Homes Grant provide funds of up to £10,000 that can be spent on solar installations.
- In several States of the United States, including California, there are tax rebates for solar energy, whether these are installed for residential, commercial or agricultural purposes. There are also rebates for solar installations for low-income families and for multi-family affordable housing projects. The exact detail of these schemes does vary between counties and states.

These are just a few examples of the schemes that are available. There are other examples and if you are considering installing a solar energy system, it is worth investigating whether there are any grant schemes or tax rebates that may help you fund part of the cost.

Financial Incentives in different countries

Financial incentives for installing solar are constantly under review, reflecting both the reducing cost of solar and the popularity of solar installations. As well as country or region-wide incentives, there are often specific incentives for different industries, particularly in the agriculture, new build and social housing sectors. It is always worth spending some time searching online to find out what incentives may be available to you.

These websites provide up-to-date details for financial incentives for different countries:

Australia	yourenergysavings.gov.au/rebates/renewable-power-incentives
Canada	renewablesassociation.ca/go-solar/

Ireland   www.seai.ie/grants/

USA   www.dsireusa.org

IN CONCLUSION

- Installing solar energy is not just good for the environment. In many cases it can make sound financial sense.
- Energy prices are increasing far faster than the general rate of inflation.
- Many countries have financial incentives to help fund solar installations, either helping with up-front capital costs, or more often by providing an ongoing income to help cover the costs of installation over the lifetime of the system.

A BRIEF INTRODUCTION TO ELECTRICITY

Before we can start playing with solar power, we need to talk about electricity. To be more precise, we need to talk about voltage, current, resistance, power and energy.

Having these terms clear in your head will help you to understand your solar system. It will also give you confidence that you are doing the right thing when it comes to designing and installing your system.

DON'T PANIC

If you have not looked at electrics since you were learning physics at school, some of the principles of electricity can be a bit daunting to start with. Do not worry if you do not fully grasp everything on your first read through.

There are a few calculations that I show on the next few pages, but I am not expecting you to remember them all! Whenever I use these calculations later in the book, I show all my workings and, of course, you can refer to this chapter as you gain more knowledge on solar energy.

Furthermore, the website that accompanies this book includes several online tools that you can use to work through most of the calculations involved in designing a solar electric system. You will not be spending hours with a slide-rule and reams of paper working all this out by yourself.

A BRIEF INTRODUCTION TO ELECTRICITY

When you think of *electricity*, what do you think of? Do you think of a battery that is storing electricity? Do you think of giant overhead pylons transporting electricity? Do you think of power stations that are generating electricity? Or do you think of a device like a kettle or television set or electric motor that is consuming electricity?

The word *electricity* covers several physical effects, all of which are related but distinct from each other. These effects are electric charge, electric current, electric potential and electromagnetism:

- An **electric charge** is a build-up of electrical energy. It is measured in coulombs. In nature, you can witness an electric charge in static electricity or in a lightning strike. A battery stores an electric charge

- An **electric current** is the flow of an electric charge, such as the flow of electricity through a cable. It is measured in amps
- An **electric potential** refers to the potential difference in electrical energy between two points, such as between the positive tip and the negative tip of a battery. It is measured in volts. The greater the electric potential (volts), the greater capacity for work the electricity has
- **Electromagnetism** is the relationship between electricity and magnetism, which enables electrical energy to be generated from mechanical energy (such as in a generator) and enables mechanical energy to be generated from electrical energy (such as in an electric motor)

HOW TO MEASURE ELECTRICITY

Voltage refers to the potential difference between two points. A good example of this is an AA battery: the voltage is the difference between the positive tip and the negative end of the battery. Voltage is measured in *volts* and has the symbol 'V'.

Current is the flow of electrons in a circuit. Current is measured in *amps* (A) and has the symbol 'I'. If you check a power supply, it will typically show the current on the supply itself.

Resistance is the opposition to an electrical current in the material the current is flowing through. Resistance is measured in *ohms* and has the symbol 'R'.

Power measures the rate of energy conversion. It is measured in *watts* (W) and has the symbol 'P'. You will see watts advertised when buying a kettle or vacuum cleaner: the higher the wattage, the more power the device consumes and the faster (hopefully) it does its job.

Energy refers to the capacity for work: power multiplied by time. Energy has the symbol 'E'. Energy is usually measured in *joules* (a joule equals one watt-second), but electrical energy is usually shown as *watt-hours* (Wh), or *kilowatt-hours* (kWh), where 1 kWh = 1,000 Wh.

VOLTS, AMPS, OHMS, WATTS AND WATT-HOURS

Volts

$$\text{Current} \times \text{Resistance} = \text{Volts}$$

$$I \times R = V$$

Voltage is equal to current multiplied by resistance. This calculation is known as Ohm's Law. As with power calculations, you can express this calculation in different ways. If you know volts and current, you can calculate resistance. If you know volts and resistance, you can calculate current:

$$\text{Volts} \div \text{Resistance} = \text{Current}$$

$$V \div R = I$$

$$\text{Volts} \div \text{Current} = \text{Resistance}$$

$$V \div I = R$$

Power

$$\text{Volts} \times \text{Current} = \text{Power}$$

$$V \times I = P$$

Power is volts multiplied by current and is measured in watts. A 12-volt circuit with a 4-amp current equals 48 watts of power (12 x 4 = 48).

Based on this calculation, we can also work out voltage if we know power and current, and current if we know voltage and power:

$$\text{Power} \div \text{Current} = \text{Volts}$$

$$P \div I = V$$

Example: A 48-watt motor with a 4-amp current is running at 12 volts.

$$48 \text{ watts} \div 4 \text{ amps} = 12 \text{ volts}$$

$$\text{Current} = \text{Power} \div \text{Volts}$$

$$I = P \div V$$

Example: a 48-watt motor with a 12-volt supply requires a 4-amp current.

$$48 \text{ watts} \div 12 \text{ volts} = 4 \text{ amps}$$

Power (watts) is also equal to the square of the current multiplied by the resistance:

$$\text{Current}^2 \times \text{Resistance} = \text{Power}$$

$$I^2 \times R = P$$

Energy

Energy is a measurement of power over time. It shows how much power is used, or generated, by a device, typically over a period of an hour. In electrical systems, it is measured in watt-hours (Wh) and kilowatt-hours (kWh).

A device that uses 50 watts of power, has an energy demand of 50Wh per hour. A solar panel that can generate 50 watts of power per hour, has an energy creation potential of 50Wh per hour.

However, because solar energy generation is so variable, based on temperature, weather conditions, the time of day and so on, a watt-peak (Wp) rating is used specifically for solar systems. A watt-peak rating shows how much power can be generated by a solar panel at its peak rating. It has been introduced to highlight the fact that the amount of energy a solar panel can generate is variable and to remind consumers that a solar panel rated at 50 watts is not going to be producing 50 watt-hours of energy every single hour of every single day.

Direct Current and Alternating Current

Two types of current can flow through an electrical circuit. Direct Current is a constant charge flowing in one direction, moving from the high voltage (positive) power source to the low voltage (negative) power source. Batteries and solar panels both work on direct currents.

An alternating current is a stream of charges that reverse direction very rapidly. The current switches directions several times each second. This cycle of switching directions is called *frequency* and is measured in *Hertz* (Hz). The faster this cycle of switching, the higher the frequency. Grid electricity works on AC power. AC power in Europe cycles around 50 times a second (50 Hz), whilst in the United States, AC power cycles 60 times a second (60 Hz).

LOW CURRENT AND HIGH CURRENT SYSTEMS

When we are designing systems, we generally want to keep the currents as low as possible. If we put too much current through a circuit, the energy loss caused by this current increase exponentially. This current creates heat and reduces the overall efficiency of the system. This resistance builds up over distance, which means that the higher the current, the more issues you will have with power loss, particularly over a long cable run.

To overcome this energy loss, you either need to install thicker and heavier cables, or increase the voltage of the system. If you double the voltage of a system, you halve the current and therefore reduce the energy loss significantly.

For example, let us say that we have two 12v, 200Wp solar panels that we wish to combine to charge a battery. We have the choice of connecting them together in a series, to create a 24v, 200-watt circuit, or connect them in parallel to create a 12v, 200-watt circuit:

- A 12-volt 100Wp solar panel has a current flow of 8.3 amps (100 watts ÷ 12-volts = 8.3 amps).
- If we connect the two solar panels up in series, the *solar array* also has a current flow of 8.3 amps (200 watts ÷ 24 volts = 8.3 amps).
- If we connect the two solar panels up in parallel, the solar array has a current flow of 16.6 amps (200 watts ÷ 12 volts = 16.6 amps).

It is usually better to double the voltage rather than double the current. Of course, there are exceptions to this rule. If, for instance, you want to use your system at 12 volts, for example, then you may decide to use thicker cables and keep cable distances to a minimum. This does limit the overall size of your system, but this may not matter. So long as you are aware of the problems and design around them, there is nothing wrong with this approach.

A WORD FOR NON-ELECTRICIANS

Realistically, if you are new to electrical systems, you should not be planning to install a big solar energy system yourself. If you want a low-voltage system to mount to the roof of a boat, garden shed or barn, or if you want to play with the technology and have some fun, then great, this book will tell you everything you need to know. However, if the limit of your electrical knowledge is wiring a plug or replacing a fuse, you should not be thinking of physically wiring and installing a solar energy system yourself without learning more about electrical systems and electrical safety first.

Furthermore, if you are planning to install a solar energy system to the roof of a house, be aware that in most countries, you need to have electrical qualifications to carry out even simple household wiring and the work you carry out may be subject to building regulations.

That does not mean that you cannot specify a solar energy system, calculate the size you need and buy the necessary hardware for a big project. It does mean that you are going to need to employ a specialist to check your design and carry out the installation.

IN CONCLUSION

- Understanding the basic rules of electricity makes it much easier to put together a solar electric system.
- As with many things in life, a bit of theory makes a lot more sense when you start applying it in practice.
- If this is your first introduction to electricity, run through it a couple of times.
- You may also find it useful to bookmark this section and refer back to it as you read on.
- You will also find that, once you have learned a bit more about solar electric systems, some of the terms and calculations will start to make a bit more sense.
- If you are not an electrician, be realistic in what you can achieve. Electrics can be dangerous, and you do not want to get it wrong. You can do most of the design work yourself, but you are going to need to get a specialist in to check your design and carry out the installation.

THE FOUR SOLAR PV CONFIGURATIONS

There are four different configurations you can choose from when creating a solar electricity installation. These are stand-alone (sometimes referred to as *off-grid*), grid-tie, grid-tie with power backup (also known as *grid interactive*) and grid fallback.

Here is a brief introduction to these different configurations:

STAND-ALONE/OFF-GRID

Until around ten years ago, stand-alone solar photovoltaic installations were the most popular type of solar installation. Solar photovoltaics were originally created for this very purpose, providing power at a location where there is no other source of power.

Whether it is powering a shed light, providing power for a pocket calculator, powering a complete off-grid home or providing electricity to an entire village in Africa or Asia, stand-alone systems fundamentally all work in the same way. The solar panel generates power, the energy is stored in a battery and then used as required.

Stand-alone systems can be a very cost-effective way of generating electricity, particularly when the overall electricity demand is comparatively low. Care must be made when designing a stand-alone system to ensure that there is enough energy production throughout the year: a system that only operates well in the summer months and fails to work in the winter is of little use!

Almost everyone can benefit from a stand-alone solar system for something, even if it is something as mundane as providing an outside light somewhere. Even if you are planning on something much bigger and grander, it is often a good idea to start with a very small and simple stand-alone system first. Learn the basics and then progress from there.

EXAMPLES OF SIMPLE STAND-ALONE SYSTEMS

Vending machines

ByBox is a manufacturer of electronic lockers, typically situated at shopping malls and fuel services and used for delivering parcels to be collected by customers.

One of the biggest issues with electronic lockers has often been finding suitable locations to place them where a power source is available. ByBox overcame this issue by building an electronic locker with a solar roof to provide permanent power to the locker.

The solar roof provides power to a set of batteries inside the locker. When not in use, the locker itself is in standby mode, thereby consuming minimal power. When a customer wishes to use the locker, they press the START button and use the locker as normal.

The benefit to ByBox has been twofold. Firstly, they can install a locker bank in any location, without any dependence on a power supply. Secondly, the cost of the solar panels and controllers is often less than the cost of installing a separate electricity supply, even if there is one nearby.

Recreational vehicles

Holidaying with recreational vehicles or caravans is on the increase, and solar energy is changing the way people are going on holiday.

In the past, most RV owners elected to stay on larger sites, which provided access to electricity and other facilities. As recreational vehicles themselves become more luxurious, however, people are now choosing to travel to more remote locations and live entirely 'off-grid', using solar energy to provide electricity wherever they happen to be. Solar is being used to provide all the comforts of home, whilst offering holidaymakers the freedom to stay wherever they want.

GRID-TIE

Grid-tie is now hugely popular in Europe, North America and Australia. This is due to the availability of grants to reduce the installation costs, the ability to earn money through feed-in tariffs and the opportunity to sell electricity back into the electricity companies, as explained in the chapter on *Making and saving money with solar* starting on page 17.

In a grid-tie system, your home runs on solar power during the day. Any surplus energy that you produce is then fed into the grid. In the evenings and at night, when your solar energy system is not producing electricity, you then buy your power from the electricity companies in the usual way.

The benefit of grid-tie solar installations is that they reduce your reliance on the big electricity companies and ensure that more of your electricity is produced in an environmentally efficient way.

One disadvantage of most grid-tie systems is that if there is a power cut, power from your solar array is also cut. By itself, you do not have additional energy security by installing a grid-tie solar energy system.

Grid-tie can work especially well in hot, sunny climates, where peak demand for electricity from the grid often coincides with the sun shining, thanks to the high power demand of air conditioning units. Grid-tie also works well where the owners use most of the power themselves.

AN EXAMPLE OF A GRID-TIE SYSTEM

Si Gelatos is a small Florida-based ice-cream manufacturer. In 2007, they installed solar panels on the roof of their factory to provide power and offset some of the energy used in running their cold storage facility.

"Running industrial freezers is extremely expensive and consumes a lot of power," explains Dan Foster of Si Gelatos. "Realistically, we could not hope to generate all our power from solar, but we felt it was important to reduce our overall power demand and solar allowed us to do that."

Cold storage facilities consume most of their power during the day in the summer, when solar is running at its peak. Since installing solar power, Si Gelatos has seen its overall energy consumption drop by 40% and now hardly takes any power from the utilities during peak operating times.

"Solar has done three things for our business," says Dan. "Firstly, it is a very visible sign for our staff that we are serious about the environment. This in turn has made our employees more aware that they need to do their bit by making sure lights and equipment are switched off when they are not needed. Secondly, it shows our customers that we care for the environment, which has been good for goodwill and sales. Thirdly, and most importantly, we're genuinely making a real contribution to the environment, by reducing our electricity demand at the time of day when everyone else's demand for electricity is high as well."

GRID-TIE WITH POWER BACKUP (GRID INTERACTIVE)

Grid-tie with power backup – also known as a *grid interactive* system – combines a grid-tie installation with a bank of batteries.

As with grid-tie, the concept is that you use power from your solar array when the sun shines and sell the surplus to the power companies. Unlike a standard grid-tie system, however, a battery bank provides contingency for power cuts so that you can continue to use power from your system.

Typically, you would set up protected circuits within your building that will continue to receive power during a power outage. This ensures that essential power remains available for running lights, refrigeration and heating controllers, for example, whilst backup power is not wasted on inessential items such as televisions and radios.

Whilst the grid is available, a grid-interactive system is configured to provide power to the whole house, with the solar and batteries providing much of the power and the grid supplementing this when required.

If there is a potential for main power to be lost for several days, it is also possible to design a system to incorporate other power generators into a grid interactive system, such as a generator. This would allow a grid interactive system to work as a highly efficient *uninterruptable power supply* (UPS) for extended periods of time.

The cost of a grid-tie system with power backup is higher than a standard grid-tie system, because of the additional cost of batteries and battery controllers. Typically, having power backup will add 35–50% of additional costs over a standard grid-tie system if using lithium batteries, or between 20–25% of additional costs if using more conventional lead acid batteries.

As with normal grid-tie systems, it is possible to sell surplus power back to the utility companies in some countries, allowing you to earn an income from your solar energy system.

There is an ever growing number of off-the-shelf grid-tie battery systems now becoming available, the Tesla Powerwall home battery being the most well-known. Most battery systems for residential use can store between four and six kilowatt-hours of energy.

AN EXAMPLE OF A GRID INTERACTIVE SYSTEM

Grid interactive systems are gaining popularity with rural farms in the United Kingdom, where even short power blackouts can cause significant disruption.

Traditionally, farms have countered this by using generators to provide light and power. However, between 2009 and 2011, when the UK Government were offering large incentives for installing solar power, many farmers fitted grid interactive systems onto their buildings, providing themselves with an income by selling electricity to the electricity utility companies and giving themselves backup power in case of a power blackout.

The additional cost of installing a grid interactive system over a standard grid-tie system is more than offset by the low running costs and ease of use of the system. Farmers do not need to buy and run generators and the system is almost entirely maintenance-free. This is a big contrast with generator systems, which need to be tested and run regularly to ensure they are working effectively.

GRID FALLBACK

With a grid fallback system, the solar array generates power which in turn charges a battery bank. Energy is taken from the battery and run through an inverter to power one or more circuits. When the batteries become depleted, the system switches back to the grid power supply until the batteries can be recharged by the solar array.

The advantage of grid-fallback systems is that they mean that specific circuits can be run entirely from solar, only switching back to the grid when the batteries are depleted. In many cases, this never happens, and you have the benefit of an off-grid system with the added confidence that the grid is there as a backup. The disadvantage of grid-fallback systems is that if the circuits are overloaded and the fallback system cannot cope with the whole load, they switch out completely so that all the power then comes from the grid until the power demand is reduced.

Grid fallback systems used to be one of my preferred solutions for small solar energy systems where overall demand for energy was low and installation costs were a primary concern. However, as costs for grid-interactive systems continue to fall, the benefits of the grid-interactive systems make grid-fallback systems less attractive.

AN EXAMPLE OF A GRID FALLBACK SYSTEM

Commercial battery storage systems for grid-tie solar installations tend to be expensive. Systems like the Tesla Powerwall allow people with solar already installed on their homes to upgrade to incorporate battery storage in their existing systems, but at a significant cost.

At the end of 2015, I was asked to design a battery storage system that was significantly cheaper than existing designs, focusing on providing customer savings through powering the lighting circuits in a home. The resulting design, called Battery LITE, stores enough energy from the solar panels each day to provide all the lighting needs for a large home.

Designed to either work with an existing grid-tie system or to be included as part of a new installation, Battery LITE allows residential solar PV customers to use their solar energy to provide free lighting both day and night on top of their existing benefits.

The system proved to be an outstanding success, saving customers significant money on their existing electricity bills. It became the best-selling solar battery system in the United Kingdom in its first year of sales.

GRID FAILOVER

You can also configure a grid fallback system as a *grid failover* system.

A grid failover system kicks in when there is a power failure from your main electricity supply. In effect, it is an uninterruptable power supply, generating its power from solar energy.

The benefit of this configuration is that if you have a power cut, you have contingency power. The disadvantage of this configuration is that you are not using solar power for your day-to-day use.

Although rare in Europe and America, grid failover systems used to be more common in countries where power failures are commonplace. In Africa and in many parts of Asia, grid failover systems reduce the reliance on power generators for lighting and basic electricity needs.

However, in most cases, customers have found that a grid fallback or grid interactive system is more suitable for their needs. I am aware of two grid failover systems that have been installed in the past. Both have since been reconfigured as grid fallback systems.

HOW GRID-TIE SYSTEMS DIFFER FROM STAND-ALONE

Generally, stand-alone and smaller grid fallback systems run at low voltages, typically between 12 and 48 volts. This is because batteries are low-voltage units and so building a stand-alone system at a low voltage is a simple, flexible and safe approach.

Grid-tie systems tend to be larger installations, often generating several kilowatts of electricity each hour. As the electricity is required as a high-voltage supply, it is more efficient to connect multiple solar panels together to produce a high voltage circuit, rather than use an inverter to step up the voltage. This high-voltage DC power is then converted into an AC current by a suitable grid-tie inverter.

Grid-tie systems either link multiple solar panels together to produce a solar array voltage of several hundred volts before running to the inverter, or have a small inverter connected to each solar panel to create a high-voltage AC supply from each panel.

The benefit of this high voltage is efficiency. There is less power loss running high-voltage, low-current electricity through cables from the solar array.

For smaller stand-alone battery-based systems, low-voltage is the best solution, as the battery banks tend to work better as low-voltage energy stores. For grid-tie systems where the energy is not being stored in a battery bank, the higher-voltage systems are the best solution. For grid-tie with battery storage, most systems create a hybrid with high-voltage power generation from the solar panels and a low or medium-voltage battery system. Neither approach is inherently 'better', it all depends on the type of system you are designing.

IN CONCLUSION

- Solar can be used in many ways and for many applications.
- Stand-alone systems are the simplest and easiest to understand. They tend to be comparatively small systems, providing power where no other power source is easily available.
- With grid-tie, your solar energy system generates electricity that is then used normally. Any excess electricity production is exported onto the grid.
- Grid-tie with power backup (also known as grid interactive) provides you with the benefits of a grid-tie system with the added benefit that power remains available even if electricity to your area is cut off.

- Grid fallback systems often have more in common with stand-alone systems than grid-tie systems. In design they are very similar to stand-alone systems, with an inverter running from a bank of batteries and an automatic transfer switch to switch power between the solar energy system and the grid power supply.
- Grid failover systems are comparatively rare now. They are most commonly used to provide uninterruptable power supplies using solar as the backup source.
- Grid-tie systems have a different design to stand-alone systems. They tend to be higher-voltage systems, whereas stand-alone systems run at much lower voltages.

COMPONENTS OF A SOLAR ELECTRIC SYSTEM

Before I get into the detail about planning and designing solar electric systems, it is worth describing all the different components of a system and explaining how they fit together. Once you have read this chapter, you will have a reasonable grasp of how a solar energy system fits together. I will not go into too much detail at this stage, simply providing an overview for now. The detail can come later.

SOLAR PANELS

The heart of a solar electric system is the solar panel itself. There are various types of solar panel, which I describe later.

Solar panels or, more accurately, *photovoltaic* solar panels, generate electricity from the sun. The more powerful the sun's energy, the more power you get, although solar panels continue to generate small amounts of electricity in the shade. Most solar panels are made up of individual solar cells connected together. A typical solar cell will only produce around half a volt, so by connecting them together in series inside the panel, a more useful voltage is achieved.

Most small solar panels – rated 150Wp or below – are rated as 12-volt solar panels, whilst larger solar panels are 24-volt panels. A 12-volt solar panel produces around 14–18 volts when put under load. This allows a single solar panel to charge up a 12-volt battery. Incidentally, if you connect a voltmeter up to a solar panel when it is not under load, you may well see voltage readings of up to 26 volts. This is normal in an 'open circuit' on a solar panel. As soon as you connect the solar panel into a circuit, this voltage level will drop to around 14–18 volts.

Solar panels can be linked together to create a *solar array*. Connecting multiple panels together allows you to produce a higher current or to run at a higher voltage:

- Connecting panels *in series* makes an array run at higher voltages. Typically, 12, 24 or 48 volts in a stand-alone system, several hundred volts in a grid-tie system
- Connecting the panels *in parallel* allows a solar array to produce more power while maintaining the same voltage as the individual panels

- When you connect multiple panels together, the power of the overall system increases, irrespective of whether they are connected in series or in parallel

In a solar array where the solar panels are connected in series (as shown in the diagram below), you add the voltages of each panel together and add the wattage of each panel together to calculate the maximum amount of power and voltage the solar array will generate.

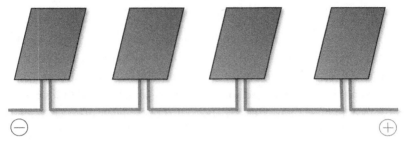

A solar array made of four solar panels connected in series. If each individual panel is rated as a 12-volt, 12-watt panel, this solar array would be rated as a 48-volt, 48-watt array with a 1 amp current.

In a solar array where the panels are connected in parallel (as shown in the diagram below), you take the *average* voltage of all the solar panels and you add the wattage of each panel to calculate the maximum amount of power the solar array will generate.

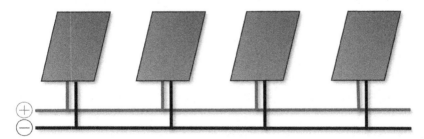

A solar array made of four solar panels connected in parallel. With each panel rated as a 12-volt, 12-watt panel, this solar array would be rated as a 12-volt, 48-watt array with a 4 amp current.

I will go into more detail later about choosing the correct voltage for your system.

BATTERIES

Solar panels rarely power electrical equipment directly. This is because the amount of power the solar array collects varies depending on the strength of the sunlight. This makes the power source too variable for most electrical equipment to cope with.

If you are storing your solar energy, this energy is stored in batteries. As well as allowing flexibility as to when you use your energy, the batteries provide a constant power source for your electrical equipment.

There are different battery technologies available for solar energy storage. Traditionally, 'deep cycle' lead acid batteries have been used. These batteries look like car batteries but have a different internal design. This design allows them to be heavily discharged and recharged several hundred times over. More recently, lithium-ion batteries have become more popular thanks to their longer life and better performance.

Most lead acid batteries are 6-volt or 12-volt batteries, whilst lithium-ion battery modules are typically 12-volt or 24-volt. Like solar panels, these can be connected to form a larger *battery bank*. Like solar panels, multiple batteries used in series increase the capacity and the voltage of a battery bank. Multiple batteries connected in parallel increase the capacity whilst keeping the voltage the same.

Lead acid battery systems have the benefit of simplicity, safety and price. They are robust and can be adapted to most applications. Lithium-ion batteries are vastly more volatile and expensive and require significantly better control to keep them safe. Typically, most lithium-ion battery systems designed for solar are pre-configured by the manufacturer to a specific capacity and performance. These systems provide a safe way of storing energy but may limit you in how you use the battery storage.

Both types of batteries are discussed in much greater detail in later chapters of the book.

CONTROLLER

If you are using batteries, your solar electric system is going to require a controller to manage the flow of electricity (the current) into and out of the battery.

If your system overcharges the batteries, this will damage and eventually destroy them. Likewise, if your system completely discharges the batteries, this will quite rapidly destroy them. A solar controller prevents this from happening.

There are a few instances where a small solar electric system does not require a controller. An example of this is a small 'battery top-up' solar panel that is used to keep a car battery in peak condition when the car is not being used. These solar panels are too small to damage the battery when the battery is fully charged.

In most other cases a solar electric system will require a controller to manage the charge and discharge of batteries and keep them in good condition.

INVERTER

The electricity generated by a solar electric system is direct current (DC). Electricity from the grid is high-voltage alternating current (AC).

If you are planning to run equipment that runs from grid-voltage electricity from your solar electric system, you will need an inverter to convert the current from DC to AC and convert the voltage to the same voltage as you get from the grid.

There are two options for an inverter: you can have one central inverter in your solar system, either connecting directly to the solar array in a grid-tie system or to the battery pack in an off-grid system. Alternatively, you can have a 'micro inverter' system where each solar panel has its own inverter and a separate inverter controller manages all the different inverters in your system to provide a harmonized high voltage AC output.

Solar panels with micro-inverters are typically only used with grid-tie systems and are not suitable for systems with battery backup. For grid-tie systems, they do offer some significant benefits over the more traditional 'big box' inverter, although the up-front cost is higher.

Inverters are a big subject all on their own. I will come back to describe them in much more detail later in the book.

ELECTRICAL DEVICES

The final element of your solar electric system are the devices you plan to power. Theoretically, anything that you can power with electricity can be powered by solar. However, many electrical devices are very power hungry, which makes running them on solar energy very expensive.

Of course, this may not be so much of an issue if you are installing a grid-tie system. If you have very energy-intensive appliances that you only use for short periods, the impact to your system is low. In comparison, running high-power appliances on an off-grid system means you must have a more powerful off-grid solar energy system to cope with the peak demand.

Low-voltage devices

Most off-grid solar systems run at low voltages. Unless you are planning a pure grid-tie installation, you may wish to consider running at least some of your devices directly from your DC supply rather than running everything through an inverter. This has the benefit of greater efficiency.

Thanks to the caravanning and boating communities, lots of equipment is available to run from a 12-volt or 24-volt supply. Light bulbs, refrigerators, ovens, kettles, toasters, coffee machines, hairdryers, vacuum cleaners, televisions, radios, air conditioning units, washing machines and laptop computers are all available to run on 12-volt or 24-volt power.

In addition, thanks to the recent uptake in solar installations, some specialist manufacturers are building ultra low-energy appliances, such as refrigerators, freezers and washing machines, specifically for people installing solar and wind turbine systems.

You can also charge up most portable items such as MP3 players and mobile phones from a 12-volt supply, either directly, or by using a step-down transformer.

Grid-voltage devices

If running everything at low voltage is not an option, or if you are using a grid-tie system, you use an inverter to run your electrical devices.

CONNECTING EVERYTHING TOGETHER

A stand-alone system

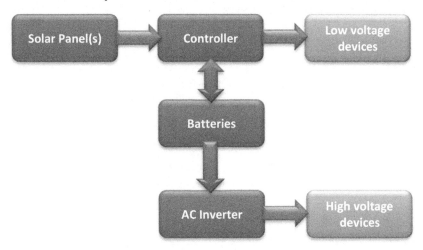

The simplified block diagram above shows a simple stand-alone solar electric system. Whilst the detail will vary, this design forms the basis of most stand-alone systems and is typical of the installations you will find in recreational vehicles, caravans, boats and buildings that do not have a conventional power supply.

This design provides both low-voltage DC power for running smaller electrical devices and appliances such as laptop computers and lighting, plus a higher-voltage AC supply for running larger devices such as larger televisions and kitchen appliances.

In this diagram, the arrows show the flow of current. The solar panels provide the energy, which is fed into the solar controller. The solar controller charges the batteries. The controller also supplies power to the low-voltage devices, using either the solar panels or the batteries as the source of this power.

The AC inverter takes its power directly from the battery and provides the grid-voltage AC power supply.

A grid-tie system using a single central inverter

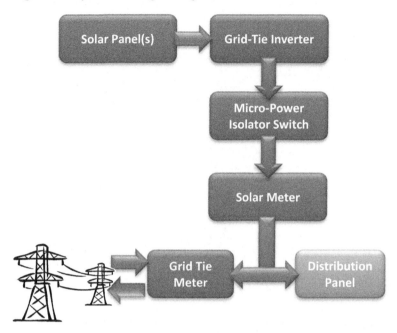

This simplified block diagram shows a simple grid-tie system, typical of the type installed in many homes today. The solar panels are connected to the grid-tie inverter, which feeds the energy into the main supply. Electricity can be used by devices in the building or fed back out onto the grid, depending on demand.

The grid-tie inverter monitors the power feed from the grid. If it detects a power cut, it also cuts power from the solar panels to ensure that no energy is fed back out onto the grid.

The solar meter records how much energy is generated by the solar panels. The grid tie meter monitors how much energy is taken from the grid. Some grid tie meters also record how much is fed back into the grid using the solar energy system.

A grid-tie system using multiple micro-inverters

A grid-tie system using micro-inverters is similar to the one above, except that each solar panel is connected to its own inverter, and the inverters themselves are daisy-chained together, converting the low-voltage DC power from each solar panel into a grid-voltage AC power supply.

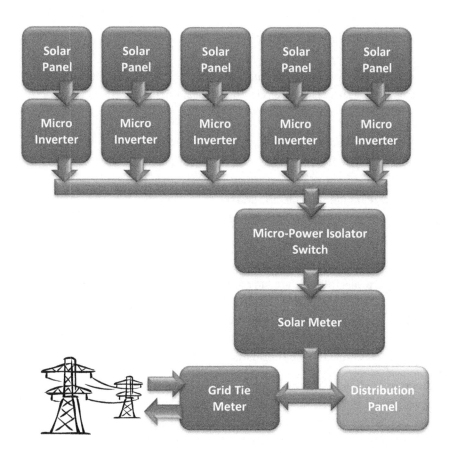

IN CONCLUSION

- There are various components that make up a solar electric system
- Multiple solar panels can be joined together to create a more powerful *solar array*.
- In a stand-alone system, the electricity is stored in batteries to provide an energy store and provide a more constant power source. A controller manages the batteries, ensuring the batteries do not get overcharged by the solar array and are not over-discharged by the devices taking current from them
- An inverter takes the DC current from the solar energy system and converts it into a grid-voltage AC current that is suitable for running devices that require grid power
- Generally, it is more efficient to use the electricity as a DC supply than an AC supply

THE DESIGN PROCESS

No matter what your solar energy system is for, there are seven steps in the design of every successful solar electric installation:

- Scope the project
- Calculate the amount of energy you need
- Calculate the amount of solar energy available
- Survey your site
- Size up the solar electric system
- Select the right components and work out full costs
- Produce the detailed design

The design process can be made more complicated, or simplified, based on the size of the project. If you are simply installing an off-the-shelf shed light, for instance, you can probably complete the whole design in around twenty minutes. If, on the other hand, you are looking to install a solar electric system in a business to provide emergency site power in the case of a power cut, your design work is likely to take considerably more time.

Whether your solar electric system is going to be large or small, whether you are buying an off-the-shelf solar lighting kit or designing something from scratch, it is worth following this basic design process every time. This is true even if you are installing an off-the-shelf system. This ensures that you will always get the best from your system and will provide you with the reassurance that your solar energy system will achieve everything you need it to do.

SHORT-CUTTING THE DESIGN WORK

Having said that doing the design work is important, there are some useful online tools to help make the process as easy as possible.

Once you have scoped your project, the Solar Electricity Handbook website, which you can find at www.SolarElectricityHandbook.com, includes several online tools and calculators that will help you carry out much of the design work.

The solar irradiance tables and solar angle calculators will allow you to work out how much solar energy is available at your location, whilst the off-grid project analysis and grid-tie project analysis questionnaires will each generate and e-mail to you a full report for your proposed system, including calculating the size of system you require and providing a cost estimate.

Of course, there is a limit to how much a set of online solar tools can help you in isolation, so you will still need to carry out a site survey and go through components selection and detailed design yourself, but these tools will allow you to try several different configurations and play out 'what if' scenarios quickly and easily.

Incidentally, whilst some of these tools ask you for an e-mail address (to send you your report), your e-mail address is not stored anywhere on the system. Other than the report that you request, you will never receive unsolicited e-mails because of entering your e-mail address.

SOLAR ENERGY AND EMOTIONS

Design can often seem to be a purely analytical and rational process. It should not be. All great designs start with a dream.

For many people, choosing solar energy is often an emotional decision. They want a solar energy system for reasons other than just the purely practical. Some people want solar energy because they want to 'do their bit' for the environment, others want the very latest technology, or want to use solar simply because it can be done. Others want solar energy because they see the opportunity to earn money. I suspect that for most people, the reasons are a combination of the above.

It is so important that the emotional reasons for wanting something are not ignored. We are not robots and our emotions should be celebrated, not suppressed. The Wright brothers built the first aircraft because they wanted to reach the sky. NASA sent a man to the moon because they wanted to go further than anyone had ever done before. Neither undertaking could be argued as purely rational; they were the results of big dreams.

It is important to acknowledge that there are often hidden reasons for wanting solar energy. Sadly, these reasons often do not make it down onto a sheet of paper in a design document or onto a computer spreadsheet. Sometimes, the person making the decision for buying solar energy is secretly worried that if they voice their dreams, they will appear in some way irrational.

The reality is that it is often a good thing if there is an emotional element to wanting a solar energy system. By documenting these reasons, you will end up with a better solution. For instance, if the environmental benefits are top of your agenda, you will use your solar energy system in a different way to somebody who is looking at solar purely as a business investment.

By acknowledging these reasons and incorporating them into the design of your system, you will end up with a far better system. Not only will you have a system that works in a practical sense, it will also achieve your dream.

IN CONCLUSION

- No matter how big or small your project, it is important to design it properly
- There are online tools available to help you with the calculations and to speed up the work
- Do not ignore the emotional reasons for wanting a solar energy system. You are a human being, you can dream!

SCOPING THE PROJECT

As with any project, before you start, you need to know what you want to achieve. In fact, it is one of the most important parts of the whole project. Get it wrong and you will end up with a system that will not do what you need it to.

It is usually best to keep your scope simple to start with. You can then flesh it out with more detail later.

Here are some examples of a suitable scope:

- To power a light and a burglar alarm in a shed on an allotment
- To provide power for lighting, a kettle, a radio and some handheld power tools in a workshop that has no conventional electrical connection
- To provide enough power for lighting, refrigeration and a TV in a holiday caravan
- To provide lighting and power to run four laptop computers and the telephone system in an office during a power cut
- To charge up an electric bike between uses
- To provide an off-grid holiday home with its entire electricity requirements
- To reduce the carbon footprint of a house or business
- To run an electric car entirely on solar energy

From your scope, you can start fleshing this out to provide some initial estimates on power requirements.

As mentioned in the previous chapter, I have created two online Solar Project Analysis tools, one for grid-tie systems and one for off-grid systems. You can find both these tools on my website www.SolarElectricityHandbook.com. You will still need to collect the basic information to work with, but all the hard work is done for you. This tool will produce a complete project scope, work out the likely performance of your solar energy system and provide some ballpark cost estimates.

For the next few chapters, I am going to use the example of providing a small off-grid holiday home with its entire electricity requirements.

This is a big project. If you have little or no experience of solar electric systems or household electrics you would be best starting with something smaller. Going completely off-grid is an ambitious project, but for the purposes of teaching solar electric system design, it is a perfect example as it requires a detailed design that covers all aspects of designing a solar energy system.

DESIGNING GRID-TIE OR GRID FALLBACK SYSTEMS

For our sample project, grid-tie is not an option as we are using solar power as an alternative to connecting our site to the electricity grid.

However, grid-tie is a popular option, especially in the United States and across Europe. This was, in part, due to the generous subsidies offered over the past few years. Whilst most of these subsidies have now come to an end or have been significantly reduced, grid-tied solar remains popular with businesses and home-owners.

In terms of scoping the project, it makes little difference whether you are planning a grid-tie or grid fallback system or not, the steps you need to go through are the same. The only exception, of course, is that you do not need to consider battery efficiencies with grid-tie.

The biggest difference with a grid-tie or grid fallback system is that you do not have to rely on your solar system providing you with all your electricity requirements as you will not be plunged into darkness if you use more electricity than you generate.

This means that you can start with a small grid-tie or grid fallback system and expand it later as funds allow.

Despite that, it is still a good idea to go through a power analysis as part of the design. Even if you do not intend to produce all the power you need with solar, having a power analysis will allow you to benchmark your system and will help you size your grid-tie system if you aim to reduce your carbon footprint by providing the electricity companies with as much power as you buy back.

Most grid-tie systems are sized to provide more power than you need during the summer and less than you need during the winter. Over a period of a year, the aim is to generate as much power as you use, although on a month-by-month basis this may not always be the case.

Many solar companies claim that this then provides you with a 'carbon neutral' system as you are selling your excess power to the electricity companies and then buying the same amount of electricity back when you need it.

If this is what you are planning to do with your grid-tie system, your scope is much simpler. You need to get your electricity bills for the past year and make a note of how much electricity you have used over the year. Then divide this figure by the number of days in the year to work out a daily energy usage and ensure your system generates this as an average over the period of a year.

Because you are not generating enough electricity during the winter months in a carbon neutral grid-tie system, you need fewer solar panels than you would need to create an entirely stand-alone system.

Comparing supply with demand

If you are designing a grid-tie system, it can be interesting to compare the supply of solar energy with your electricity usage pattern. By comparing supply with demand, you can see how closely solar energy production matches your own usage and this, in turn, can be used as an indicator to identify how environmentally beneficial solar energy is for you.

To do this, you will need to ascertain your monthly electricity usage for each month of the year. Your electricity supplier may already provide you with this information on your electricity bill. If not, you should be able to request this from them.

Once you have this information, visit *www.SolarElectricityHandbook.com* and fill in the Grid-Tie Solar Project Analysis, including your individual monthly consumption figures. In the report that is e-mailed to you, you will see a chart that allows you to see how closely your electricity usage maps onto solar energy production.

This report will also provide you with an approximate estimate for the carbon footprint for each kilowatt-hour of electricity you produce from your solar array, based on the production and installation of your solar array and the likely amount of energy that it will generate during its lifetime.

Based on this, it is possible to see whether installing solar energy is likely to produce *real-world* environmental savings.

FLESHING OUT THE SCOPE

Now we know the outline scope of our project, we need to quantify exactly what we need to achieve and work out some estimates for energy consumption.

Our holiday home is a small two-bedroom cottage with a solid fuel cooker and boiler. The cost of connecting the cottage to the grid is £4,100 (around $6,520) and I suspect that solar electric power could work out significantly cheaper.

The cottage is mainly used in the spring, summer and autumn, with only a few weekend visits during the winter.

Electricity is required for lighting in each room, plus a courtesy light in the porch, a fridge in the kitchen and a small television in the sitting room. There also needs to be surplus electricity for charging up a mobile phone or MP3 player and for the occasional use of a laptop computer.

Now we have decided what devices we need to power, we need to find out how much energy each device needs and estimate the daily usage of each item.

To keep efficiency high and costs low, we are going to work with low-voltage electrics wherever possible. The benefits of using low-volt devices rather than higher grid-voltage are twofold:

- We are not losing efficiency by converting low-volt DC electrics to grid-voltage AC electrics through an inverter.
- Many electronic devices that plug into a grid-voltage socket require a transformer to reduce the power back down to a low DC current, thereby creating a second level of inefficiency

Many household devices, like smaller televisions, music systems, computer games consoles and laptop computers, have external transformers. It is possible to buy transformers that run on 12-volt or 24-volt electrics rather than the AC voltages we get from the grid. This is the most efficient way of providing low-voltage power to these devices.

There can be disadvantages of low-voltage configurations, however, and they are not the right approach for every project:

- If running everything at 12–24 volts requires a significant amount of additional rewiring, the cost of carrying out the rewiring can be much higher than the cost of an inverter and a slightly larger solar array
- If the cable running between your batteries and your devices is too long, you will get greater power losses through the cable at lower voltages than you will at higher voltages

If you already have wiring in place to work at grid-level voltages, it is often more appropriate to run a system at grid voltage using an inverter, rather than running the whole system at low voltage. If you have no wiring in place, running the system at 12 or 24 volts is often more suitable.

PRODUCING A POWER ANALYSIS

The next step is to investigate your power requirements by carrying out a power analysis, where you measure your power consumption in watt-hours.

You can find out the wattage of household appliances in one of four ways:

- Check the rear of the appliance, or on the power supply
- Check the product manual
- Measure the watts using a watt meter
- Find a ballpark figure for similar items

Often a power supply will show an output current in amps rather than the number of watts the device consumes. If the power supply also shows the output voltage, you can work out the wattage by multiplying the voltage by the current (amps):

$$\text{Power }(watts) = \text{Volts} \times \text{Current }(amps)$$
$$P = V \times I$$

For example, if you have a mobile phone charger that uses 1.2 amps at 5 volts, you can multiply 1.2 amps by 5 volts to work out the number of watts. In this example, it equals 6 watts of power. If I plugged this charger in for one hour, I would use 6 watt-hours of energy.

A watt meter, like the example shown on the right, is a useful tool for measuring the energy requirements of any device that runs on high-voltage AC power from the grid. The watt meter plugs into the wall socket and the appliance plugs into the watt meter. An LCD display on the watt meter then displays the amount of power the device is using. This is the most accurate way of measuring your true power consumption.

Finding a ballpark figure for similar devices is the least accurate way of finding out the power requirement and should only be done as a last resort. A list of power ratings for some common household appliances is included in Appendix A.

Once you have a list of the power requirements for each electrical device, draw up a table listing each device, whether the device uses 12-volt or grid voltage, and the power requirement in watts.

Then estimate how long you will use each device each day and multiply the watts by hours to create a total watt-hour energy requirement for each item.

You should also factor in any 'phantom loads' on the system. A phantom load is the name given to devices that use power even when they are switched off. Televisions in standby mode are one such example, but any device that has a power supply built into the plug also has a phantom load. These items should be unplugged or switched off at the switch when not in use. However, you may wish to factor in a small amount of power for items in standby mode, to factor in the times you forget to switch something off.

If you have a gas-powered central heating system, remember that most central heating systems have an electric pump and the central heating controller will require electricity as well. A typical central heating pump uses between 25 and 60 watts of power, which can easily add up to 500-800 watt-hours per day, whilst a central heating controller can use between 2 and 24 watt-hours a day.

Once complete, your power analysis will look like this:

Device	Voltage	Power (watts)	Hours of use per day	Watt-hours energy
Living room lighting	12V	11W	5	55Wh
Kitchen lighting	12V	11W	2	22Wh
Hallway lighting	12V	8W	½	4Wh
Bathroom lighting	12V	11W	1½	17Wh
Bedroom 1 lighting	12V	11W	1	11Wh
Bedroom 2 lighting	12V	11W	1½	17Wh
Porch light	12V	8W	½	4Wh
Small fridge	12V	12W	24	288Wh
TV	12V	40W	4	160Wh
Laptop computer	12V	40W	1	40Wh
Charging cellphones	12V	5W	4	20Wh
Phantom loads	12V	1W	24	24Wh
Total Energy Requirement a day (watt-hours)				**662Wh**

A word of warning

In the headlong enthusiasm for implementing a solar electric system, it is very easy to underestimate the amount of electricity you need at this stage.

To be sure that you do not leave something out which you regret later, I suggest you have a break at this point. Then return and review your power analysis.

It can help to show this list to somebody else to get their input as well. It is very easy to get emotionally involved in your solar project. Having a second pair of eyes can make a world of difference to make sure you are not overlooking something.

WHEN YOU ARE READY TO PROCEED

We now know exactly how much energy we need to store to provide one day of power. For our holiday home example, that equates to 662 watt-hours per day.

We still need to take the efficiency of the overall system into account. Batteries, inverters and resistance in the circuits all reduce the efficiency of our solar electric system. We must consider these inefficiencies and add them to our power analysis.

CALCULATING INEFFICIENCIES

Batteries do not return 100% of the energy used to charge them. The *Charge Cycle Efficiency* of the battery measures the percentage of energy available from the battery compared to the amount of energy used to charge it.

Charge cycle efficiency is not a fixed figure, as the efficiency can vary depending on how quickly you charge and discharge the battery. However, most solar applications do not overstress batteries and so the standard charge cycle efficiency figures are usually sufficient.

Approximate charge cycle efficiency figures are normally available from the battery manufacturers. However, for industrial quality 'traction' batteries, you can assume 95% efficiency, whilst gel batteries and leisure batteries are usually in the region of 90%.

If you are using an inverter in your system, you need to factor in the inefficiencies of the inverter. Again, the actual figures should be available from the manufacturer but typically, you will find that an inverter is around 90% efficient.

Adding the inefficiencies to our power analysis

In our holiday home example, there is no inverter. If there were, we would need to add 10% for inverter inefficiencies for every grid-powered device. We are using batteries. We need to add 5% to the total energy requirement to take charge cycle efficiency into account. 5% of 662 equals 33 watts. Add this to our power analysis, and our total watt-hour requirement becomes 695 watt-hours per day.

WHEN DO YOU NEED TO USE THE SOLAR SYSTEM?

It is important to work out at what times of year you will be using your solar electric system most. For instance, if you are planning to use your system full time during the depths of winter, your solar electric system needs to be able to provide all your electricity even during the dull days of winter.

A holiday home is often in regular use during the spring, summer and autumn, but left empty for periods of time during the winter. This means that, during winter, we do not need our solar electric system to provide enough electricity for full occupancy. We need enough capacity in the batteries to provide enough electricity for, say, the occasional long weekend. The solar array can then recharge the batteries again, once the home is empty.

We might also decide that, if we needed additional electricity in winter, we could have a small standby generator on hand to give the batteries a boost charge.

For the purposes of our holiday home, our system must provide enough electricity for full occupancy from March to October and occasional weekend use from November until February.

KEEPING IT SIMPLE

You have seen what needs to be considered when creating a power analysis and calculating the inefficiencies. If you are planning to use the online tools to help you, now is the time to use them.

Visit *www.SolarElectricityHandbook.com* and follow the links to either the Off-Grid or Grid-Tied Solar Project Analysis tools, which can be found in the *Online Calculators* section. This will allow you to enter your devices on the power analysis, select the months you want your system to work and select your location from a worldwide list. The system will automatically e-mail you a detailed solar analysis report with all the calculations worked out for you.

IMPROVING THE SCOPE

Based on the work done, it is time to put more detail on our original scope. Originally, our scope looked like this:

Provide an off-grid holiday home with its entire electricity requirements.

Now the improved scope has become:

Provide an off-grid holiday home with its entire electricity requirements, providing power for lighting, refrigeration, TV, laptop computer and various sundries, which equals 695 watt-hours of electricity consumption per day.

The system must provide enough power for occupation from March until October, plus occasional weekend use during the winter.

There is now a focus for the project. We know what we need to achieve for a solar electric system to work. Now we need to go to the site and see if what we want to do is achievable.

IN CONCLUSION

- Getting the scope right is important for the whole project
- Start by keeping it simple and then flesh it out by calculating the energy requirements for all the devices you need to power
- If you are designing a grid-tie system, you do not need to go into so much detail, you can usually find the usage information on your electricity bill
- If you are designing a grid-tie system, you can make a reasonable estimate of its environmental benefit by comparing solar energy supply with your demand on a month-by-month basis
- Remember to factor in 'phantom loads'
- Because solar electric systems run at low voltages, running your devices at low voltage is more efficient than inverting the voltage to grid levels first
- Thanks to the popularity of caravans and boats, there is a large selection of 12-volt appliances available. If you are planning a stand-alone or grid fallback system, you may wish to use these in your solar electric system rather than less efficient grid-voltage appliances
- Remember to factor in inefficiencies for batteries and inverters
- Consider the times of year that you need to use your solar electric system
- Once you have completed this stage, you will know what the project needs to achieve to be successful

CALCULATING SOLAR ENERGY

The next two chapters are just as useful for people wishing to install a solar hot water system as they are for people wishing to install solar electricity.

Whenever I refer to *solar panel* or *solar array* (multiple solar electric panels) in these two chapters, the information is equally valid for solar electricity and solar hot water.

WHAT IS SOLAR ENERGY?

Solar energy is a combination of the hours of sunlight you get at your site and the strength of that sunlight. This varies depending on the time of year and where you live.

This combination of hours and strength of sunlight is called *solar insolation* or *solar irradiance*, and the results can be expressed as watts per square metre (W/m^2) or, more usefully, in kilowatt-hours per square metre spread over the period of a day ($kWh/m^2/day$). One square metre is equal to 9.9 square feet.

Why is this useful?

Photovoltaic solar panels quote the expected number of watts of power they can generate, based on a solar irradiance of 1,000 watts per square metre. This figure is often shown as a watts-peak (Wp) figure and shows how much power the solar panel can produce in ideal conditions. A solar irradiance of 1,000 watts per square metre is what you could expect to receive at solar noon in the middle of summer at the equator. It is not an average reading that you could expect to achieve on a daily basis.

However, once you know the solar irradiance for your area, quoted as a daily average (i.e. the number of kilowatt-hours per square metre per day), you can multiply this figure by the wattage of the solar panel to give you an idea of the daily amount of energy you can expect your solar panels to provide.

Calculating solar irradiance

Solar irradiance varies significantly from one place to another and changes throughout the year. To come up with some reasonable estimates, you need irradiance figures for each month of the year for your specific location.

Thanks to NASA, calculating your own solar irradiance is simple. NASA's network of weather satellites has been monitoring the solar irradiance across the surface of the earth for many decades. Their figures take into account the upper atmospheric conditions, average cloud cover and surface temperature, and are based on sample readings every three hours for the past quarter of a decade. They cover the entire globe.

The Solar Electricity Handbook website has incorporated this information across the entire world. Simply select your current location and you can view the irradiance figures for your exact area. You can access this solar irradiance information by visiting http://www.SolarElectricityHandbook.com and following the links to calculators and resources.

Using the information from the website, here are the solar irradiance figures for London in the United Kingdom, shown on a month-by-month basis. They show the average daily irradiance, based on mounting the solar array flat on the ground:

	Jan	Feb	Mar	Apr	May	Jun	Jul	Aug	Sep	Oct	Nov	Dec
Flat	0.75	1.37	2.31	3.57	4.59	4.86	4.82	4.20	2.81	1.69	0.92	0.60

These figures show how many hours of equivalent midday sun we get over the period of an average day of each month. In the chart above, you can see that in December we get the equivalent of 0.6 of an hour of midday sun (36 minutes), whilst in June we get the equivalent of 4.86 hours of midday sunlight (4 hours and 50 minutes).

Capturing more of the sun's energy

The tilt of a solar panel has an impact on how much sunlight you capture: mount the solar panel flat against a wall or flat on the ground and you will capture less sunlight throughout the day than if you tilt the solar panels to face the sun.

The figures above show the solar irradiance in London, based on the amount of sunlight shining on a single square metre of the ground. If you mount your solar panel at an angle, tilted towards the sun, you can capture more sunlight and therefore generate more power. This is especially true in the winter months, when the sun is low in the sky.

The reason for this is the intensity of the sunlight. When the sun is high in the sky the intensity of sunlight is high. When the sun is low in the sky the sunlight is spread over a greater surface area:

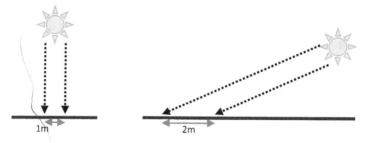

This diagram shows the different intensity of light depending on the angle of sun in the sky. When the sun is directly overhead, a 1m-wide shaft of sunlight will cover a 1m-wide area on the ground. When the sun is low in the sky – in this example, I am using an angle of 30° towards the sun – a 1m-wide shaft of sunlight will cover a 2m-wide area on the ground. This means the intensity of the sunlight is half as much when the sun is at an angle of 30° compared to the intensity of the sunlight when the sun is directly overhead.

The impact of tilting solar panels on solar irradiance

If we tilt our solar panels towards the sun, it means we can capture more of the sun's energy to convert into electricity. Often the angle of this tilt is determined for you by the angle of an existing roof. However, for every location there are optimal angles at which to mount your solar array, to capture as much solar energy as possible.

Using London as an example again, this chart shows the difference in performance of solar panels, based on the angle at which they are mounted. The angles I have shown are flat on the ground, upright against a wall, and mounted at different angles designed to get the optimal amount of solar irradiance at different times of the year (I explain the relevance of these specific angles in a moment):

Jan Feb Mar Apr May Jun Jul Aug Sep Oct Nov Dec

FLAT

0.75 1.37 2.31 3.57 4.59 4.86 4.82 4.20 2.81 1.69 0.92 0.60

UPRIGHT

1.20 1.80 2.18 2.58 2.70 2.62 2.71 2.80 2.47 2.07 1.43 1.01

BEST ALL YEAR ROUND PERFORMANCE – 38° TILT

1.27 2.04 2.76 3.67 4.17 4.20 4.25 4.16 3.26 2.41 1.53 1.05

BEST WINTER PERFORMANCE – 23° TILT

1.30 2.03 2.62 3.34 3.66 3.69 3.76 3.73 3.06 2.37 1.56 1.08

BEST SUMMER PERFORMANCE – 53° TILT

1.19 1.95 2.77 3.84 4.52 4.63 4.66 4.41 3.31 2.33 1.43 0.97

ADJUSTING THE TILT THROUGHOUT THE YEAR FOR OPTIMAL PERFORMANCE

1.30 2.05 2.78 3.86 4.70 4.91 4.90 4.46 3.31 2.41 1.56 1.08
22°   30°   38°   46°   54°   62°   54°   46°   38°   30°   22°   14°
tilt tilt tilt tilt tilt tilt tilt tilt tilt tilt tilt tilt

Note: All angles are given in degrees from vertical and are location specific.

Look at the difference in the performance based on the tilt of the solar panel. Look at the difference in performance in the depths of winter and in the height of summer.

It is easy to see that some angles provide better performance in winter; others provide better performance in summer, whilst others provide a good compromise all-year-round solution.

Calculating the optimum tilt for solar panels

Because of the 23½° tilt of the earth relative to the sun, the optimum tilt of your solar panels will vary throughout the year, depending on the season. In some installations, it is feasible to adjust the tilt of the solar panels each month, whilst in others it is necessary to have the array fixed in position.

To calculate the optimum tilt of your solar panels, you can use the following sum:

90° − your latitude = optimum fixed year-round setting

This angle is the optimum tilt for fixed solar panels for all-year-round power generation. This does *not* mean that you will get the maximum power output every single month, instead it means that across the whole year, this tilt will give you the best compromise, generating electricity all the year round.

Getting the best from solar panels at different times of the year

Depending on when you want to use your solar energy, you may choose to use a different tilt to improve power output at a given point in the year. Each month of the year, the angle of the sun in the sky changes by 7.8° – higher in the summer and lower in the winter. By adjusting the tilt of your solar panel to track the sun, you can tweak the performance of your system according to your requirements.

You may want to do this for several reasons:

- For a stand alone, off-grid system, you often need to get as much power generation during the winter months as possible to counter the reduction in natural light
- When installing a grid-tie system in a cool climate, where the focus is on reducing your carbon footprint, you can choose to boost your electricity production in winter, to offset the amount of electricity you need to buy when demand is at its highest
- When installing a grid-tie system where the focus is on making the most profit by selling your power, you can choose to tilt your solar panels at a summer setting and thereby produce the maximum amount of energy possible over the course of the year

You can see this monthly optimum angle (rounded to the nearest whole degree) on the bottom row of the table on the previous page.

Optimum winter settings

Here is an example of how you could tweak your system. Performance of a solar system is at its worst during the winter months. However, by tilting your panels to capture as much of the sunlight as possible during the winter, you can significantly boost the amount of power you generate at this time.

Based on the Northern Hemisphere, an optimum winter tilt for solar panels is the optimum angle for November and January. For the Southern Hemisphere, the optimum winter tilt is May and July:

$$90° - \text{your latitude} - 15.6° = \text{optimum winter setting}$$

As you can see from the table on the previous page, if you tilt your solar panels at this angle, you will sacrifice some of your power generation capability during the summer months. However, as you are generating so much more power during the summer than you are in the winter, this may not be an issue.

More importantly, compared to leaving the panels on a flat surface, you are almost doubling the amount of power you can generate during the three bleakest months of the year. This means you can reduce the number of solar panels you need to install.

Optimum summer settings

If you wish to get the best output of your system overall, you will find that you will get slightly more energy, when measured over the course of the whole year, by angling your solar panels to an optimum summer time tilt.

In warm climates, where maximum energy consumption is during hot weather, angling your panels to get the maximum amount of sunlight during the height of summer can be the best solution, both financially and environmentally.

For Northern Hemisphere countries, the optimum summer time tilt is the optimum angle for May and July. For Southern Hemisphere countries, the optimum summer time tilt is the optimum angle for November and January:

$$90° - \text{your latitude} + 15.6° = \text{optimum summer setting}$$

Positioning your solar panels

Regardless of where you live, the sun always rises from the east and sets in the west. If you live in the Northern Hemisphere, solar panels will always work best if they are south-facing. In the Southern Hemisphere, solar panels work best if they are north-facing.

However, it is not always possible to position your solar panels so they are aligned perfectly to point towards the midday sun. For instance, if you want to install solar panels on the roof of the house, and the roof faces east/west, then your solar panels will inevitably be less efficient.

Thanks to improved solar panel design over the past decade, this is not as big a problem as was once the case. Whilst the figure varies slightly in different parts of the world, from one solar panel manufacturer to another and during the different seasons of the year, the average efficiency drop of a solar panel mounted away from due south (due north in the Southern Hemisphere) is around 1.1% for every five degrees.

This means that if your panels face due east or due west, you can expect around 20% loss of efficiency compared to facing your panels in the optimum position.

If you face your panels in the completely opposite direction – north in the Northern Hemisphere or south in the Southern Hemisphere – losing around 40% of the efficiency of your solar array.

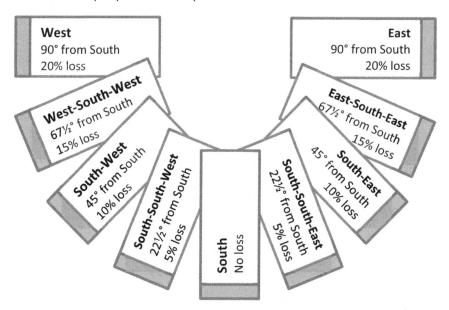

For the Northern Hemisphere, this chart shows the approximate efficiency loss by not facing your panels directly south. For Southern Hemisphere countries, the chart should be reversed.

Using solar irradiance to work out how much energy a solar panel will generate

Based on these figures, we can calculate on a monthly basis how much power a solar panel will give us per day, by multiplying the monthly solar irradiance figure by the stated wattage of the panel:

Solar Irradiance x Panel Wattage (the solar panel watt-peak rating) = Watt-hours per day

As we now know, the solar irradiance figure depends on the month and the angle for the solar panel. Assuming we have a 20Wp solar panel, mounted flat on the ground, here are the calculations for London in December and June:

	December	June
Flat	0.60 x 20W = 12 Wh of energy per day	4.86 x 20W = 97 Wh of energy per day

As you can see, there is a big difference in the amount of energy you can generate in the middle of summer, compared to winter. In the example above, over eight times the amount of energy is generated in the height of summer compared to the depths of winter.

Here are the same calculations again, but with the solar panels angled at 38° for best all-year-round performance. Note the significant improvement in winter performance and the slightly reduced summer performance:

	December	June
38° angle	1.05 x 20W = 21 Wh of energy per day	4.20 x 20W = 84 Wh of energy per day

Using solar irradiance to give you an approximate guide for the required power capacity of your solar array

In the same way that you can work out how much energy a solar panel will generate per day, you can use solar irradiance to give you an approximate guide for the required capacity of solar array that you need.

I say an approximate guide, because the *actual* capacity will also need to take into account:

- The peculiarities of your site
- The location and angles of your solar panels
- Any obstacles blocking the sunlight at different times of year

I cover all this in the next chapter when I look at the site survey. Nevertheless, it can be useful to carry out this calculation to establish a ballpark cost for your solar electric system. The calculation is simple. Take the figure you calculated for your total number of watt-hours per day and divide it by the solar irradiance figure for the worst month that you require your system to work.

Using our holiday home as an example, we can look at our watt-hours per day figure of 695Wh/day and then divide this number by the worst month on our irradiance chart (December). It is worth doing this based on mounting the solar panel at different angles, to see how the performances compare:

Flat	695 ÷ 0.6 = 1159 watts	If we have our solar panels laid flat, we would need a 1,159Wp solar array to power our home in December.
Upright	695 ÷ 1.01 = 688 watts	Mounted vertically on a wall, the solar panels could generate the same amount of power with a 688Wp solar array.
38° angle	695 ÷ 1.05 = 661 watts	Angled towards the equator, we could generate the same amount of power with a 661Wp solar array.
23° angle Best winter tilt	695 ÷ 1.08 = 643 watts	With the optimum winter tilt, we can us a 643Wp solar array.
53° angle Best summer	695 ÷ 0.97 = 716 watts	Angled towards the summer sun, we would require a 716Wp solar array to provide power in December
Adjusted Monthly	695 ÷ 1.08 = 643 watts	With the tilt of the solar panel adjusted each month, we can use a 643Wp solar array, the same as the best winter tilt settings.

This chart tells us that to provide full power for our holiday home in December, we require a solar array with a generation capacity of between 643 watts and 1159 watts, depending on the tilt of the solar panels.

But remember our scope. We only want to use the home full time from March to October. The solar electric system only needs to provide enough electricity for a long weekend during the winter. This means that, so long as our batteries are big enough to provide electrical power for a few days, it does not matter if the solar power in winter is not enough to provide for constant use. As soon as we close up the holiday home again, the solar panels will recharge the batteries.

Here are my calculations again, this time using October as our worst month:

Flat	695 ÷ 1.69 = 411 watts	If we have our solar panels laid flat, we would need a 411Wp solar array to power our home in October.
Upright	695 ÷ 2.07 = 335 watts	If we mount the solar panels vertically, we could generate the same amount of power with a 335Wp solar array.
38° angle	695 ÷ 2.41 = 288 watts	Angled towards the equator, we could generate the same amount of power with a 288Wp solar array.
23° angle Best winter tilt	695 ÷ 2.37 = 293 watts	With the optimum winter tilt, we can use a 293Wp solar array.
53° angle Best summer	695 ÷ 2.33 = 298 watts	Angled towards the summer sun, we would require a 298Wp solar array to provide power in October

	695 ÷ 2.41 = 288 watts	With the tilt of the solar panel adjusted each month, we can use a 288Wp solar array, the same as the best year-round tilt settings.
Adjusted Monthly		

This chart tells us that we require a solar array with a generation capacity of between 288Wp and 411Wp, depending on the tilt of the solar panels. Compared to our earlier calculations for generating power throughout the year, it is much lower. We have just saved ourselves a significant amount of money.

You can also see that during the summer months, the solar electric system will generate considerably *more* electricity than we will need to run our holiday home. That is fine. Too much is better than not enough and it allows for the occasions when a light is left switched on or a TV is left on standby.

SOLAR PANELS AND SHADE

The biggest negative impact on solar energy production is shade. Even if only a very small amount of your solar array is in shade, the impact on the performance of your whole system can have a very big effect.

Unlike solar thermal (hot water) systems, the loss of power through shading is much greater than the amount of the panel that is in shade. With solar thermal systems, if 5% of the panel is in shade, you lose around 5% of the power production.

Depending on the exact circumstances, even if only 5% of a photovoltaic solar panel is in shade, it is possible to lose 50–80% of power production from your entire solar array.

For this reason, it is hugely important that your solar energy system remains out of shade throughout the day wherever possible. Sometimes this is not possible, and this requires some additional design work to keep the effect of shade on your system to a minimum.

I cover shade in much more detail in the next chapter, including an explanation of why shade has such a big impact on energy production. For now, it is just important to know that shading can significantly affect the amount of energy you can get from your solar energy system.

SOLAR ARRAY POWER POINT EFFICIENCIES

Now we know the theoretical size of our solar panels. However, we have not taken into account the efficiencies of the panels themselves or the efficiencies of the controller or inverter that handles them.

Solar panels are rated on their 'peak power output'. Peak power on a solar panel in bright sunlight is normally generated at between 14 and 20 volts. This voltage can go up and down quite significantly, depending on the amount of sunlight available.

This swing in voltage gets much higher if you have multiple solar panels connected together in series – or if you are using a higher-voltage solar panel. It is common for a solar array with fifteen or twenty panels connected in series to have voltage swings of several hundred volts when a cloud obscures the sun for a few seconds!

Managing this voltage swing can be done in one of two ways. The cheap and simple method is to cut the voltage down to a setting that the solar panel can easily maintain. For instance, a solar panel rated at 12 volts will usually maintain a voltage level of at least 14 volts during daylight hours. A charge controller or inverter that cuts the voltage down to this level will then always be able to use this power. The disadvantage of this approach is that, as you cut the voltage, the wattage drops with it, meaning you can lose a significant amount of energy.

In terms of the amount of energy you can capture, as opposed to what the solar array collects, this method can reduce the efficiency of a solar array by around 25%.

A better solution is to use controllers and inverters that incorporate *Maximum Power Point Tracking* (MPPT). Maximum power point tracking adjusts the voltage from the solar array to provide the correct voltage for the batteries or for the inverter to remove this inefficiency.

Maximum power point trackers are typically 90–95% efficient. Over the past five years, the price of MPPT controllers and inverters has dropped and the availability has increased to the point where it is almost always worth buying an MPPT controller and inverter for all but the very smallest and simplest solar installations.

Incidentally, you only need an MPPT inverter for grid-tie systems where you are powering the inverter directly from the solar panels. You do not require an MPPT inverter if you are planning to run the inverter through a battery bank.

To account for power point efficiencies, you need to divide your calculation by 0.9 if you plan to use a MPPT controller or inverter, and by 0.75 if you plan to use a non-MPPT controller or inverter:

Non MPPT controller/inverter calculation		MPPT controller/inverter calculations	
Flat solar panel	Solar panel at 38° tilt	Flat solar panel	Solar panel at 38° tilt
411 watts ÷ 0.75 =	288 watts ÷ 0.75 =	411 watts ÷ 0.9 =	288 watts ÷ 0.9 =
548Wp solar panel	384Wp solar panel	456Wp solar panel	320Wp solar panel

THE EFFECTS OF TEMPERATURE ON SOLAR PANELS

Solar panels will generate less power when exposed to high temperatures compared to when they are in a cooler climate. Solar PV systems can often generate more electricity on a day with a cool wind and a hazy sun than when the sun is blazing and the temperature is high.

When solar panels are given a wattage rating, they are tested at 25°C (77°F) against a 1,000 W/m² light source. At a cooler temperature, the solar panel will generate more electricity, whilst at a warmer temperature the same solar panel will generate less.

As solar panels are exposed to the sun, they heat up, mainly due to the infrared radiation they are absorbing. As solar panels are dark, they can heat up quite considerably. In a hot climate, a solar panel can quite easily heat up to 80–90°C (160–175°F).

Solar panel manufacturers provide information to show the effects of temperature on their panels. Called a *temperature coefficient of power* rating, it is shown as a percentage of total power reduction per 1°C increase in temperature.

Typically, this figure will be in the region of 0.5%, which means that for every 1°C increase in temperature, you will lose 0.5% efficiency from your solar array, whilst for every 1°C decrease in temperature you will improve the efficiency of your solar array by 0.5%.

Assuming a temperature coefficient of power rating of 0.5%, this is the impact on performance for a 100W solar panel at different temperatures:

	5°c / 41°F	15°c / 59°F	25°c / 77°F	35°c / 95°F	45°c / 113°F	55°c / 131°F	65°c / 149°F	75°c / 167°F	85°c / 185°F
Panel output	110W	105W	100W	95W	90W	85W	80W	75W	70W
% gain/ loss	10%	5%	0%	-5%	-10%	-15%	-20%	-25%	-30%

In northern Europe and Canada, high temperature is not a significant factor when designing a solar system. However, in southern states of America and in Africa, India, Australia and the Middle East, where temperatures are significantly above 25°C (77°F) for much of the year, the temperature of the solar panels may be an important factor when planning your system.

If you are designing a system for all-year-round use, then in all fairness a slight dip in performance at the height of summer is probably not an issue for you. If that is the case, you do not need to consider temperature within the design of your system and you can skip the information on the next page.

If your ambient temperature is high during the times of year you need to get maximum performance from your solar panels, then you will need to account for temperature in your design.

You can help reduce the temperature of your solar panels by ensuring a free flow of air both above and below the panels. If you are planning to mount your solar panels on a roof, make sure there is a gap of around 7–10cm (3–4") below the panels, to allow a flow of air around them. Alternatively, you can consider mounting the panels on a pole, which will also aid cooling.

For a roof-top installation, if the average air temperature at a particular time of year is 25°C/77°F or above, multiply this temperature *in Celsius* by 1.4 to get a likely solar panel temperature. For a pole-mounted installation, multiply your air temperature by 1.2 to calculate the likely solar panel temperature. Then increase your wattage requirements by the percentage loss shown in the *temperature coefficient of power* rating shown on your solar panels, to work out the wattage you need your solar panels to generate.

Temperature impact on solar performance in Austin, Texas during the summer months

By way of an example, here is a table for Austin in Texas. This shows average air temperatures for each month of the year, the estimated solar panel temperature for the hottest months of the year and the impact on the performance on the solar array, assuming a temperature coefficient of power rating of 0.5%.

	Jan	Feb	Mar	Apr	May	Jun	Jul	Aug	Sep	Oct	Nov	Dec
Ave month temp	49°F / 10°C	53°F / 12°C	62°F / 16°C	70°F / 21°C	76°F / 24°C	81°F / 27°C	85°F / 29°C	85°F / 29°C	81°F / 27°C	71°F / 22°C	61°F / 16°C	52°F / 11°C
Likely temp of roof-mounted solar array *(Temp x 1.4)*					93°F / 34°C	100°F / 38°C	106°F / 41°C	106°F / 41°C	100°F / 34°C			
Performance impact for roof-mounted solar					-4½%	-6½%	-8%	-8%	-6½%			
Likely pole-mounted temperature of solar array *(Temp x 1.2)*					84°F / 29°C	90°F / 32°C	95°F / 35°C	95°F / 35°C	90°F / 32°C			
Performance for pole-mounted solar					-2%	-3½%	-5%	-5%	-3½%			

The *Performance Impact* calculations in rows 3 and 5 of the above table are calculated using the following formula:

(Estimated Solar Panel Temp °C − 25°C) x (−Temp Coefficient of Power Rating)

So, for July, the calculation for roof-mounted solar was (41°C − 25°C) x (−0.5) = − 8%

For around five months of the year, the ambient temperature in Austin is greater than 25°C/77°F. During these months, the higher temperature will mean lower power output from a solar array. If you are designing a system that must operate at maximum efficiency during the height of summer, you will need to increase the size of your solar array by the percentages shown, to handle this performance decrease.

You can find the average ambient air temperature for your location by visiting *The Weather Channel* website at *www.weather.com*. This excellent site provides average monthly temperatures for towns and cities across the world, shown in your choice of Fahrenheit or Celsius.

Our example holiday home project is in the United Kingdom, where the temperature is below 25°C for most of the year. In addition, our solar design will produce more power than we need during the summer months. As a result, we can ignore temperature in our particular project.

WORKING OUT AN APPROXIMATE COST

It is worth stressing again that these figures are only approximate at this stage. We have not yet taken into account the site itself and we are assuming that shading will not be an issue.

If you are planning to do the physical installation yourself, a solar electric system consisting of solar array, controller and battery costs around £1.50–£2.50 ($2–3.50 US) per watt, +/− 15%.

A grid-tie system is cheaper to install than a stand-alone system, as you do not need to budget for batteries. You will, however, need a qualified electrician to certify the system before use. In most countries, you will also need all the components used in your solar installation to be certified as suitable for grid installation. If you are planning grid-tie, budget around £0.80–£1.50 ($1–$2 US) per watt, +/− 15%.

For our holiday home installation, we need 320 watts of solar electricity if we tilt the solar panels towards the sun, or 456 watts if we mount the panels flat. Our rough estimate suggests a total system cost of between £500–800 ($650–1050 US), +/– 15% for tilted panels, or £685–1150 ($890–1500 US) +/– 15% for a flat panel installation.

If you remember, the cost to connect this holiday home to a conventional electricity supply was £4,500 ($7,200). Therefore, installing solar energy is the cheaper option for providing electricity for our home.

What if the figures do not add up?

In some installations, you will get to this stage and find out that a solar electric system simply is unaffordable. This is not uncommon. I was once asked to calculate the viability for using 100% solar energy at an industrial unit once, and came up with a ballpark figure of £33½m (around $54m)!

When this happens, you can do one of two things: walk away, or go back to your original scope and see what can be changed.

The best thing to do is go back to the original power analysis and see what savings you can make. Look at the efficiencies of the equipment you are using and see if you can make savings by using lower-energy equipment or changing the way equipment is used.

If you are absolutely determined to implement a solar electric system, there is usually a way to do it. However, you may need to be ruthless as to what you leave out.

In the example of the industrial unit, the underlying requirement was to provide emergency lighting and power for a cluster of computer servers if there was a power cut. The cost for implementing this system was around £32,500 ($52,000), which was comparable in cost to installing and maintaining on-site emergency generators and UPS equipment.

WORKING OUT DIMENSIONS

Now we know the capacity of the solar panels, we can work out an approximate size of our solar array. This is extremely useful information to know before we carry out our site survey as the solar panels must go somewhere. We need to be able to find enough suitable space for them where they will receive uninterrupted sunshine in a safe location.

There are two main technologies of solar panels on the market, *amorphous* and *crystalline*. I will explain the characteristics and the advantages and disadvantages of each later.

For the purposes of working out how much space you're going to need to fit the solar panels, you need to know that a 1m² (approximately 9.9 square feet) amorphous solar panel generates in the region of 60 watts, whilst a 1m² crystalline solar panel generates in the region of 160 watts.

Therefore, for our holiday home, we are looking for a location where we can fit between 5 and 7.6m² (49–75 square feet) of amorphous solar panels or 2–3m² (20–29 square feet) of crystalline solar panels.

IN CONCLUSION

- By calculating the amount of solar energy *theoretically* available at our site, we can calculate ballpark costs for our solar electric system
- There are various inefficiencies that must be considered when planning your system. If you do not take these into account, your system may not generate enough power
- It is not unusual for these ballpark costs to be far too expensive on your first attempt. The answer is to look closely at your original scope and see what can be cut to produce a cost-effective solution
- As well as working out ballpark costs, these calculations also help us work out the approximate dimensions of the solar array. This means we know how much space we need to find when we are carrying out a site survey

SURVEYING YOUR SITE

The site survey is one of the most important aspects of designing a successful solar system. It will identify if your site is suitable for solar. If it is, the survey identifies the ideal position to install your system, ensuring that you get the best value for money and the best possible performance.

WHAT WE WANT TO ACHIEVE

For a solar electric system to work well, we need the site survey to answer two questions:

- Is there anywhere on the site that is suitable for positioning my solar array?
- Do nearby obstacles such as trees and buildings shade out too much sunlight?

The first question might sound daft but, depending on your project, it can make the difference between a solar energy system being viable or not.

By answering the second question, you can identify how much of the available sunlight you will be able to capture. It is vitally important that you answer this question. The number one reason for solar energy failing to reach expectations is obstacles blocking out sunlight, which dramatically reduces the efficiency of the system.

To answer this second question, we need to be able to plot the position of the sun through the sky at different times of the year. During the winter, the sun is much lower in the sky than it is during the summer months. It is important to ensure that the solar array can receive direct sunlight throughout the day during the winter.

What you will need

You will need a compass, a protractor, a spirit level and a tape measure.

Inevitably a ladder is required if you are planning to mount the solar array on a roof. A camera can also be extremely useful for photographing the site. If you have an iPhone or an Android cell phone, you can also download some cheap software that will help you identify the path of the sun across the sky and assist with obstacle analysis.

I also find it useful to get some large cardboard boxes. Open them out and cut them into the rough size of your proposed solar array. This can help you when finding a location for your solar panels. It is far easier to envisage what the installation will look like and it can help highlight any installation issues that you might otherwise have missed.

If you have never done a solar site survey before, it does help if you visit the site on a sunny day.

Once you have more experience with doing solar site surveys, you will find it does not actually make much difference whether you do your site survey on a sunny day or an overcast day. As part of the site survey, we manually plot the sun's position across the sky, so once you have more experience, sunny weather makes little difference to the quality of the survey.

FIRST IMPRESSIONS

When you first arrive on the site, the first thing to check is that the layout of the site gives it access to sunlight.

We will use a more scientific approach for checking for shade later, but a quick look first often highlights problems without needing to carry out a more in-depth survey.

If you are in the Northern Hemisphere, look from east, through south and to the west to ensure there are no obvious obstructions that can block the sunshine, such as trees and other buildings. If you are in the Southern Hemisphere, you need to check from east, through north to west for obstructions. If you are standing on the equator, the sun passes overhead, so only obstructions in the east and west are important.

Be very careful not to look directly at the sun, even for a few moments, whilst you are carrying out this survey. Even in the middle of winter, retina burn can cause permanent damage to your eyesight.

Look around the site and identify all the different options for positioning the solar array. If you are considering mounting your solar array on a roof, remember that the world looks a very different place from a roof-top, and obstructions that are a problem standing on the ground look very different when you are at roof height.

Drawing a rough sketch of the site

It can be helpful to draw up a rough sketch of the site. It does not have to be exact, but it can be a useful tool to have, both during the site survey and afterwards when you are designing your system.

Include all properties and trees that are close to your site and not just those on your land. Include trees that are too small to worry about now but may become a problem in a few years' time. Also make a note of which way is north.

POSITIONING THE SOLAR ARRAY

Your next task is to identify the best location to position your solar array. Whilst you may already have a good idea where you want to install your solar panels, it is always a good idea to consider all the different options available to you.

As we discovered in the last chapter, solar arrays perform at their best when tilted towards the sun.

If you are planning to install solar energy for a building, then the roof of the building can often be a suitable place to install the solar array. This is effective where the roof is south-facing or where the roof is flat and you can fit the panels using angled mountings.

Other alternatives are to mount solar panels on a wall. This can work well with longer, slimmer panels that can be mounted at an angle without protruding too far out from the wall itself. Alternatively, solar panels can be ground-mounted or mounted on a pole.

When considering a position for your solar array, you need to consider how easy it is going to be to clean the solar panels. Solar panels do not need to be spotless, but dirt and grime will reduce the efficiency of your solar system over time, so while you are looking at mounting solutions it is definitely worth considering how you can access your panels to give them a quick wash every few months.

Roof-mounting

If you are planning to mount your solar array on a roof, you need to gain access to the roof to check its suitability.

Use a compass to check the direction of the roof. You will always get the best performance if your roof is south facing (in the northern hemisphere), but you can still get reasonable performance even if your roof is east or west facing.

You will also need to find out the pitch of the roof. Professionals use a tool called a *Roof Angle Finder* to calculate this. Roof angle finders (also called *Magnetic Polycast Protractors*) are low-cost tools available from builders' merchants. You press the angle finder up against the rafters underneath the roof and the angle finder will show the angle in degrees.

Alternatively, you can calculate the angle using a protractor at the base of a roof rafter underneath the roof or download an angle finder onto your mobile phone and use that instead.

Solar panels in themselves are not heavy: 15–20 kilograms (33–44 pounds) at most. Yet when multiple panels are combined with a frame, especially if that frame is angled, the overall weight can become quite significant. A typical 4kWp grid-tie system, made up of sixteen 250Wp solar panels can often weigh in the region of 360kg (almost 800 pounds).

Check the structure of the roof. Ensure that it is strong enough to take the solar array and to ascertain what fixings you will need. It is difficult to provide general advice on roof structures and fixings. There are so many different roof designs it is not possible for me to provide much useful information on this subject. If you are not certain about the suitability of your roof, ask a builder to assess your roof for you.

Shading issues on roofs

Sometimes, roofs can have shading issues of their own. Chimneys, television aerials and satellite dishes or vents in the roof can either cause problems with fitting the solar panels or can cast shade across the solar panels at different times of the day. Sometimes, these problems are easily resolved, moving an aerial is a relatively straightforward job, for example. However, if there is a chimney on the roof that is likely to cast a shadow across your solar panels, you will need to ensure enough space between your panels and the chimney to ensure that shadows are kept to an absolute minimum.

Measuring the available roof space

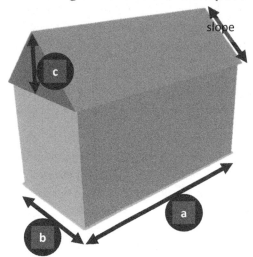

You will need to measure and record the overall roof-space available for your solar array. There are various ways of doing this, but the easiest way is to measure the footprint of the building at ground floor level, measure the height of the roof from within the attic space and then calculate the size of the roof from this.

If your roof looks like the one on the left, you have a gabled roof, which is the easiest type to measure.

From ground floor level, measure the length of the house (a) and the depth of the house (b).

Now measure the height of the roof, from the bottom of the roof to the top (c). This is usually easiest to do from within the building, going up into the attic space.

You can now calculate the slope of the roof, using the following the following formula:

$$slope = \sqrt{c^2 \times (\tfrac{1}{2} b)^2}$$

It is recommended that you either use inches or cm for your measurements. For example, if your house is 1000cm long and 500cm deep, and the height of your roofspace is 250cm, type this into your scientific calculator:

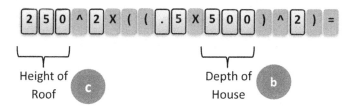

In this example, the answer is 353.5cm. This means the slope of your roof (measurement d in the diagram at the top of the page) is 353.5cm (139 inches).

You now need to deduct some of this space. It is never advisable to fit solar panels right to the edge of a roof. In some countries, such as the United Kingdom, there is a minimum space around the panels that must be left.

The reason for this is wind loading. The effects of wind speed and pressure on roof-mounted solar panels can mean that solar panels mounted too close to the edge of a roof can either cause damage to the solar panels or to the roof in extreme weather conditions. Consequently, it is widely recommended that solar panels are not fitted within 30cm (12 inches) of any roof edge. In the United Kingdom, this has been written into building regulations and are a legal requirement for professional installations.

Therefore, you should now subtract 60cm (24 inches) from both the length of the building *(a)* and the slope.

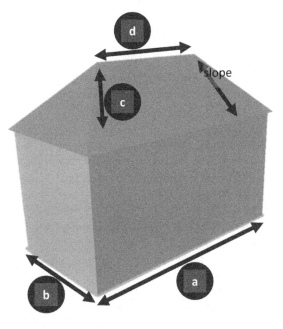

You now know the amount of roof space you have available: the length of the building *(a)* multiplied by the slope of the roof. So in the example above, the total available roof space is 9940cm long by 293.5cm high.

If you have a hip roof, like the one shown on the right, your calculation will be a little more complicated. A hip roof is a type of roof where all the sides slope downwards to the walls. An absolutely accurate calculation to work out precisely the amount of space available with a hipped roof can get very convoluted. However, there is a quick and simple ready-reckoner calculation that is accurate enough unless you are trying to squeeze every last solar panel onto the roof.

Measure the length and the width of the building (measurements *a* and *b*) from ground floor level, but as well as measuring the height of the roof from within the attic, you also need to measure the length of the roof between the hip ends (measurement *d*).

You then calculate the slope of the roof in the same way as you would for a gable roof, as shown on the previous page.

To calculate the overall length of the roof available, you then do two calculations, one for the roof between the hip ends (measurement *d*), one for the hip ends of the building and then add them both together.

In this example, if the house is 1000cm long *(a)* and 500cm deep *(b)*, the roof height is 250cm *(c)* and the length between the hips is 500cm *(d)*:

- Calculate your slope using the equation on the previous page (slope = 353.5cm). Then subtract 60cm – or 24 inches if using inches – from this figure to take into account that you cannot fit solar panels too close to either the top or the bottom of the roof (353.5 – 60 = 293.5cm).
- Calculate the roof space between the hips (500cm x 293.5cm)
- Subtract *(d)* from *(a)* to work out the overall length of the hip ends, then subtract 30cm to take into account that you cannot fit solar panels too close to the edge of the roof.
- Multiply this figure by half the slope (470cm x 146.75cm)

This will give you an approximate size that you can use for working out how many solar panels will then fit on your roof.

In some cases, you will have a pyramid roof, or may wish to fit solar panels onto the hip-end of a roof, in which case you have a fully triangular roof to work with. In this scenario, it is best to do your measurements inside the attic.

Again, the calculations for working out how many solar panels will fit on this type of roof can get fairly convoluted, but again, there is an easy calculation which is used by many installers for working out how many solar panels can be fitted.

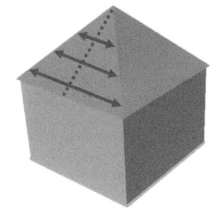

Measure 120cm (48 inches) down from the top of the roof along the pitch of the roof, and measure the width of the roof at this point. Deduct 60cm (24 inches) from this measurement to take into account that you cannot fit solar panels too close to the edge of the roof.

Then measure a further 100cm (40 inches) down the pitch of the roof and measure the width of the roof at this second point. Again, deduct 60cm (24 inches) from this measurement.

Continue measuring at 100cm (40 inch) intervals down to the bottom of the roof.

You then have a good idea of how many rows of solar panels you can fit in and the width of each row.

Ground-mounting

If you want to mount your solar array on the ground, you will need a frame onto which you can mount your solar panels. Most solar panel suppliers can supply suitable frames or you can fabricate your own on site.

There are benefits for a ground-mounted solar array. You can easily keep the panels clean and you can use a frame to change the angle of the array at different times of year to track the height of the sun in the sky.

Take a note of ground conditions, as you will need to build foundations for your frame.

Incidentally, you can buy ground-mounted solar frames that can also move the panel to track the sun across the sky during the day. These *Solar Trackers* can increase the amount of sunlight captured by around 15–20% in winter and up to 55% in summer.

Unfortunately, at present, commercial solar trackers are expensive. Unless space is at an absolute premium, you would be better to spend your money on a bigger solar array.

However, for a keen DIY engineer who likes the idea of a challenge, a solar tracker that moves the array to face the sun as it moves across the sky during the day could be a useful and interesting project to do. There are various sites on the internet, such as *instructables.com*, where keen hobbyists have built their own solar trackers and provide instructions on how to make them.

Pole-mounting

Another option for mounting a solar panel is to affix one on a pole. Because of the weight and size of the solar panel, you will need an extremely good foundation and a heavyweight pole, to withstand the wind.

You can mount up to 600-watt arrays using single-pole mountings. Larger arrays can be pole-mounted using frames constructed using two or four poles.

Most suppliers of solar panels and associated equipment can provide suitable poles.

Splitting the solar array into several smaller arrays

It may be that when you get to the site, you find that there is no one space available that will allow you to install all the solar panels you need. If this is the case, it is possible to split your single solar array into several smaller arrays. This means, for instance, that you could have two sets of panels mounted on different roof pitches, or some mounted on a roof and some from a pole.

If you do this, you are creating two *separate* solar energy systems, which you then have to link together. For a grid-tie system, you would require either a micro-inverter system or an inverter that can accept inputs from more than one solar array. For a stand-alone system, you would require a battery controller for each separate solar array.

IDENTIFYING THE PATH OF THE SUN ACROSS THE SKY

Once you have identified a suitable position for your solar array, it is time to be a little more scientific in ensuring there are no obstructions that will block sunlight at different times of the day, or at certain times of the year.

The path of the sun across the sky changes throughout the year. This is why carrying out a site survey is so important as you cannot just check to see where the sun is shining today. The height and position of the sun constantly changes throughout the year.

Each year, there are two days in the year when the day is exactly twelve hours long. These two days are 21st March and 21st September, the *solar equinoxes*. On these equinoxes the sun rises due east of the equator and sets due west of the equator. At solar noon on the equinox, exactly six hours after the sun has risen, the angle of the sun is 90° minus the local latitude.

In the Northern Hemisphere (i.e. north of the equator), the longest day of the year is 21st June and the shortest day of the year is 21st December. These two days are the summer and winter solstices respectively.

On the summer solstice, the angle of the sun is 23.5° higher than it is on the equinox, whilst on the winter solstice the angle is 23.5° lower than on the equinox.

These two extremes are due to the tilt of the earth, relative to its orbit around the sun. In the Northern Hemisphere, the summer solstice occurs when the North Pole is tilted towards the sun, and the winter solstice occurs when the North Pole is tilted away from the sun.

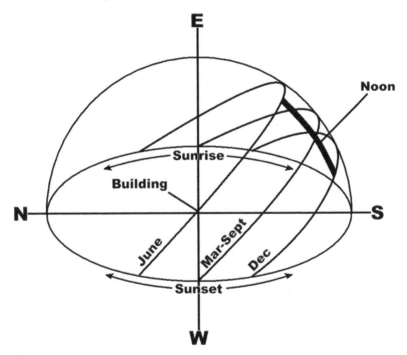

Figure 4: This chart shows the different paths of the sun from sunrise to sunset at different times of the year from the Northern Hemisphere. The intersection between N, S, E and W is your location.

We will take London in the United Kingdom as an example. London's latitude is 51°. On the equinox, the angle of the sun at noon will be 39° (90° − 51°). On the summer solstice, the angle will be 62.5° (39° + 23.5°) and on the winter solstice, it will be 15.5° (39° − 23.5°).

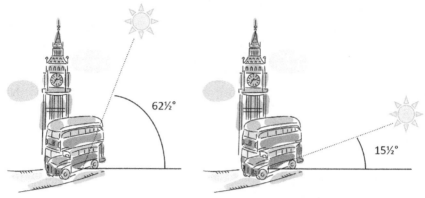

London in mid-summer   London in mid-winter

For more detailed information on sun heights on a monthly basis, or for information for other countries, visit *www.SolarElectricityHandbook.com* and follow the link to the solar panel angle calculator.

SHADING

As mentioned in the previous chapter, shading can have a very significant impact on the performance of your solar energy system. Even a tiny amount of shading can have a huge impact on the amount of energy that your system is able to generate. Therefore, it is important that your solar array remains shade-free whenever possible.

Being entirely shade free is often impossible to achieve. If so, reducing shade to an absolute minimum by carefully evaluating your site is important.

Why shading is a problem

Unlike solar thermal (hot water) systems, the loss of power through shading is much greater than the amount of the panel that is in shade. With solar thermal systems, if only 5% of the panel is in shade, you lose less than 5% of the power production. As mentioned earlier, the difference is significantly more with photovoltaic panels. Shade can bring power generation down by as much as 80%, even if the amount of the panel in shade is small.

The reason for this is in the construction of the solar panel itself. A crystalline solar panel is made up of individual solar cells. Typically, each of these cells generates ½ volt of potential energy. These cells are connected in series to increase the voltage to a more useful level inside the solar panel. Each individual solar panel has one or more strings of solar cells.

Because these strings of cells are connected in series, a solar panel is only as good as its weakest cell. If one cell produces a weak output, all the other cells within the string are compromised as well. This means that if you have a 'soft shade' such as a distant tree branch, creating a soft dappled shade across just one or two cells on a solar panel, the effect can be to reduce the output of the whole string to a similar level to a dull, overcast day.

Worse happens when you have a more direct shadow, creating a bigger contrast between light and shade. When even one cell is in complete shade and the remainder are in bright sunlight, the shaded cell short-circuits as the flow of electrons within the cell goes into reverse.

However, because the cell is connected in series with the other cells in the panel, current is forced through the reversed cell. In this scenario, the reversed cell *absorbs* the power produced by the other cells in the panel, generating heat and creating a hotspot within the solar panel.

The amount of power that is absorbed by a single reversed cell is disproportionate to the amount of power a single cell can generate. A single cell may only produce ½ volt of potential energy, but can absorb 6–8 volts when it has gone into reversal.

If unchecked, a solar cell left in this state can easily be destroyed. The hot spot generated by the blocked cell can very quickly reach dangerous temperature levels too, and amateur home-built solar panels have been known to burst into flames precisely for this reason.

Thankfully, solar panel manufacturers have a solution to this problem to avoid the panel itself becoming damaged. All professionally manufactured crystalline solar panels have in- built protection to route power around a string of cells where one or more of the cells are in reversal.

If your solar panel has only a single string of solar cells within it, this effectively means the entire solar panel is bypassed and produces no power at all. If your solar panel has two strings of solar cells, it means the power output of the panel is halved.

Because a solar array is typically put together by installing multiple solar panels in series, the effects of shading on just part of one panel impacts the performance of the entire solar array.

Analysing shade

When analysing shade, there are two different types of shading to be concerned with. Hard shading refers to an object, such as a building or tree within 10 metres (33 feet) of your proposed location that creates shade. Soft shading refers to an object beyond 10 metres of your location. Soft shading from objects that are further away still allow diffused light to reach the solar panel and have less of an effect on performance throughout the day.

You can carry out this analysis in various ways. You can use a sun path diagram with compass and pencil, a professional obstacle analysis tool or download a solar shading application onto your mobile phone or tablet.

Using a sun path diagram with compass

This is the simplest way to identify shading issues with solar. Used properly, it is an accurate way to identify shading issues with your proposed system.

A sun path diagram shows the location of the sun at different hours of the day for different months of the year. The vertical axis shows the height of the sun in the sky, whilst the horizontal axis shows the direction of the sun (the azimuth). The diagram can be used to identify when nearby objects will cast a shadow across a solar panel.

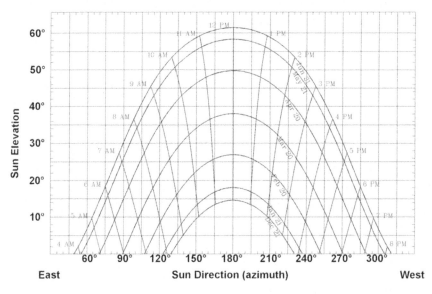

Above: A sun chart diagram, produced by the University of Oregon Solar Radiation Monitoring Laboratory.

On the chart above, the red lines depict the time of the day, whilst the blue lines show the path of the sun across the sky for different times of the year. The blue line path at the top of the chart shows the path of the sun at the height of summer (the summer solstice), whilst the blue line path at the bottom of the chart shows the path of the sun at the depths of winter (the winter solstice).

There are a number of sources for sun charts. In the United Kingdom, the Microgeneration Certificate Scheme publish a sun chart that must be used if you are planning to install a system that can take advantage of the UK Feed In Tariff scheme. Alternatively, you can download world wide sun charts from the University of Oregon Solar Radiation Monitoring Laboratory web site at http://solardat.uoregon.edu/SunChartProgram.php.

Once you have got your sun chart, you will require a compass and a clinometer. A clinometer is a tool for measuring elevation angles. If you do not have one, you can either download one onto a smartphone or tablet, or make one using a protractor, a pencil and some tape.

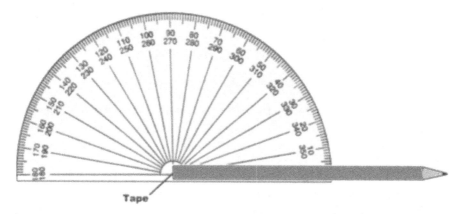

Tape the pencil to the centre of the protractor where all the lines meet, in such a way that the other end of the pencil can be moved across the protractor. You can use this protractor to check the field of view, using the pencil as an 'aimer' to measure the angle of nearby obstacles relative to your location.

Now stand at the location where you are planning to install your solar panels. Use a compass to identify magnetic south (magnetic North in the Southern Hemisphere). Then working from east to west, identify any obstacles within 90% of south (i.e. between 90° and 270° on your compass. Measure the angle of the top of each obstacle using your clinometer, and measure the width of the obstacle in degrees using your compass.

Now mark the height and width of the obstacle on your sun chart and block off this area on the chart. For example, if a nearby building is between 120–150° on the compass and has a 15° elevation, you would block off this area on the chart (shown in yellow, below):

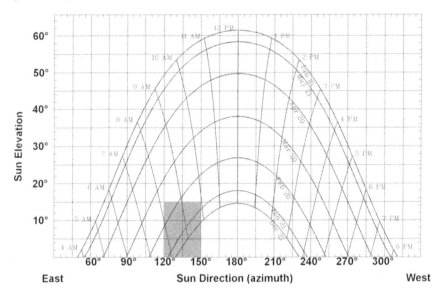

You may well have more than one obstacle to mark off in this way.

If the obstacle creating the shade is within ten metres (33 feet) of the proposed location, you then need to draw a *shade circle* around the obstacle on the chart to reflect the more severe impact that these items will have on the solar panel performance.

In the example above, if the building creating the shade issue is more than ten metres away, we need do nothing more with the chart. However, if the building is closer, we must draw a circle from the bottom line of the chart to the height of the obstacle, based from the centre of the obstacle:

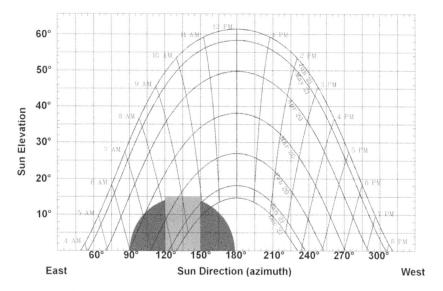

Once you have completed marking in all the shade, you must count all the squares that have been blocked out by the obstacles. If only a partial square is covered, include it in your count.

Ignore any squares below the blue line path marking the winter solstice, or any squares above the blue line path for the summer solstice. Also ignore any squares more than 90° away from south (or north, if you are in the Southern Hemisphere):

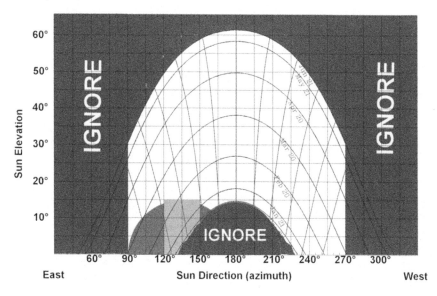

Based on this, you can see that our one obstacle either fully or partially covers twelve squares. Each square that you count is equivalent to 1% of lost solar irradiation. So for our chosen location, we can calculate that there is approximately a 12% energy loss due to shading.

As a rough rule of thumb, if you find a location with no shading whatsoever, you are doing very well. Anything less than 10% energy loss is usually regarded as very good, whilst any location with more than 25% shading can mean significant extra costs involved in building your solar energy system.

Professional tools for obstacle analysis

The paper and pencil approach I have just described works extremely well, but requires a bit of practice and can be time consuming. If you are planning on doing several solar installations, investing in a professional obstacle analysis tool or phone app is money worth spending.

In the past, a product called a Solar Pathfinder was one of the best tools you could get. This was a plastic unit with an angle chart mounted on the top. A glass dome was then placed on top of the chart. You would mount the unit onto a tripod at the desired location. Obstacles were reflected in the glass bubble and this would allow you to manually plot the obstacles on the chart and then manually work out your shading issues.

The Solar Pathfinder is surprisingly effective, as it can be easily moved around to find the best location for solar panels. Many professional solar installers still use a Solar Pathfinder for quick checks, despite also using the more expensive and advanced electronic equipment to provide a more detailed analysis.

Some solar suppliers can rent you a Solar Pathfinder for a small daily or weekly fee and can do the manual calculations for you once you have plotted the obstacles.

Today there are electronic systems that use GPS, tilt switches and accelerometers to do this work electronically. They are expensive to buy or rent on a daily basis. They do provide extremely comprehensive solar analysis, however, and if you plan to take up solar installations professionally, they are a worthwhile investment.

The best known is the *Asset* from Wiley Electronics and the *SunEye* from Solmetric. My personal preference is *SunEye*, as I find the unit simpler, but both products do a similar job.

Mobile phone and tablet applications

Whilst solar pathfinders are very good, these tools are overkill for smaller solar installations and are often very complicated to use. Thankfully, modern smart phones and tablets provide the processing power and functionality to do a similar job. A few companies have now developed solar shade analysis software to run on these mobile phones. These use the phone's built-in GPS, compass, accelerometer and camera to record a complete shade analysis in a matter of a few moments.

Some of these applications are free, whilst others cost a small amount of money. Two products that I have used are Solmetric IPV, running on iPhone, and Solar Shading, running on Android devices. Solmetric IPV costs $29.99 and handles your obstacle tracking, automatically providing charts showing your shading analysis throughout the year. The detail of reporting is not as great as some of the other electronic tools, but more and more professionals are now using this software. It provides most of the functions that you get with a more expensive system, but in a package that is easier to use and far cheaper to buy.

You can find out more about *Solmetric IPV* from www.solmetric.com and the software can be downloaded from the iTunes website.

Solar Shading is produced by Comoving Magnetics. Costing $15 on the Android Market, this application provides you with a complete shading analysis throughout the year, presenting the information in easy-to-read charts.

Considering the future

You do need to consider the future when installing a solar electric system. The system will have a lifetime of at least 20 years, so as far as possible you need to ensure that the system will be effective for that length of time.

When scanning the horizon, remember trees and hedges will grow during the lifetime of the system. A small spruce in a nearby garden could grow into a monster in the space of a few years, and if that is a risk, it is best to know about it now, rather than have a nasty surprise a few years down the line.

See if there is any planned building work nearby that may overshadow your site and try to assess the likelihood of future building work that could have an impact on shading.

It is also worth finding out if fog or heavy mist is a problem at certain times of the year. The efficiency of your solar array will be compromised if the site has regular problems with heavy mist.

What if there are shading obstructions?

Once you have identified any shading obstructions, you can design your system around them. There are various ways of countering shading issues.

Unless you have significant shading issues, you can often design around the problems using one of the options described below, or by using a combination of options to come up with a workable solution. Over the past few years, I have designed many shade-tolerant systems, including systems that are designed to work in an entirely shaded environment all year round.

Increasing the number of solar panels

If you have the space, the first option is to increase the number of solar panels you install to counter the periods of shade. Thanks to the falling price of solar panels, this can often be the cheapest solution to the problem.

This is not always possible, either because of space or cost restrictions. Also, it is not always the most efficient way of getting around the problem.

Panel orientation

If shade affects you at a particular time of the day, consider angling your solar panels away from the obstruction. This will increase their effectiveness during the unobstructed parts of the day and reduce or remove the impact of the obstruction in the first place. The reduction in power generation from angling the panels away from due south is often less than the impact on shading if you can eliminate the shade problem altogether.

Choice of solar panel

Another option is to choose amorphous (thin-film) solar panels. Amorphous solar panels do not suffer from cell reversal in the same way that crystalline solar panels do, and consequently provide far better shade tolerance.

Because of their lower efficiency levels, you will need around twice as much physical space to install amorphous solar panels. If space is not an issue, using amorphous solar panels is likely to be the simplest and most cost-effective solution for *stand-alone* solar installations.

Sharp, Mitsubishi, Uni-Solar, Solar Frontier and Sanyo now manufacture high-quality amorphous solar panels that offer excellent performance and reliability.

Use micro-inverters

If you are building a grid-tie system, possibly the best solution is to use micro-inverters, where each panel has its own power inverter and is a full solar energy system in its own right.

With a micro-inverter system, each solar panel runs entirely independently of every other solar panel. If shade affects one panel, none of the other panels is affected in any way.

Using micro-inverters also means that you can have solar panels facing in different directions from each other. For instance, if you are installing solar panels on a roof and your roof has multiple pitches and angles, you can choose to install your solar array on more than one part of the roof using a micro-inverter system.

This approach gives you greater flexibility as to where you mount solar panels. It may even be possible to avoid your shading issues altogether.

Design a multi-string solar array

Instead of designing your system to have just one set of solar panels connected in series, you can design your system to have multiple series of solar panels. Controllers and inverters are available that allow you to have multiple strings of solar panels, or you can have a controller or inverter for each individual string.

In effect, this means you end up with two smaller solar power systems, rather than one. This may mean you can face your two different sets of solar arrays at different angles, giving you the opportunity to mount them in entirely different locations if you so wish.

In this scenario, you can design your system so that partial shading will only affect one of your strings rather than your entire array. As with micro-inverters, such an approach may even allow you to avoid the shading issue altogether.

If all else fails…

Sometimes it is not possible, or not cost effective, to design around shading problems. In this scenario, it requires a rethink of what you can achieve using solar.

This is an issue that I had with my previous home. My house was on the edge of ancient woodland. For most of the year, my home was almost entirely shaded by the tall trees that surround it. Even in the height of summer, the sun did not reach my garden until mid-afternoon.

This shading meant I could never run my own home completely from solar power. However, I designed a smaller solar energy system that provided me with enough energy to run my lighting and provided backup power in case of a power cut.

I achieved this by installing amorphous solar panels onto the back of my home, facing south-west to catch the afternoon and evening sun. Despite receiving very little sunlight, this system provided enough electricity to provide me with my lighting and backup power requirements throughout the year, even in the depths of winter.

POSITIONING BATTERIES, CONTROLLERS AND INVERTERS

You need to identify a suitable location for batteries, controllers and inverters within your system. This could be a room within a building, or in an attic space, in a garage or garden shed, or in a weatherproof housing.

It is important to try to keep all the hardware close together, to keep the cable lengths as short as possible. By 'hardware', I am referring to the solar array itself, batteries, controller and inverter.

For the batteries, inverter and controller, you are looking for a location that fits the following criteria:

- Water- and weather proof
- Not affected by direct sunlight
- Insulated to protect against extremes of temperature
- Facilities to ventilate gases from the batteries
- Protected from sources of ignition
- Away from children, pets and rodents

Lead acid batteries give off very small quantities of hydrogen when charging. Hydrogen is explosive. You must ensure that wherever your batteries are stored, the area receives adequate external ventilation to ensure these gases cannot build up. You should also consider fitting a hydrogen alarm to warn you of any build-up of hydrogen over a period of time.

Lithium-ion batteries are very susceptible to extremes of heat and may explode or catch fire if overheated. For this reason, lithium-ion batteries are usually installed on an inside wall in a garage or inside a home where heat build up is much less of a problem. Lithium ion batteries should never be installed in an attic space.

Because of the extremely high potential currents involved solar PV, the ancillary equipment must be in a secure area away from children and pets.

This ancillary equipment can be quite heavy. A residential inverter will typically weigh between 15-20kg (33-44 pounds), whilst a residential battery controller will weigh around 10-15kg (22-33 pounds). A 4kWh lithium-ion battery system will weigh between 60-100kg (132-220 pounds), whilst a similar lead acid battery system will weigh around 120kg (265 pounds).

Most commonly, residential sized controllers and inverters are wall mounted, often in a garage or inside a loft, whilst industrial scale systems are floor mounted. Batteries are often mounted on heavy duty racking, or in a purpose built battery enclosure.

CABLING

While you are on site, consider likely routes for cables, especially the heavy-duty cables that link the solar array, controller, batteries and inverter together. Try to keep cable lengths as short as possible, as longer cables mean lower efficiency. Measure the lengths of these cables so that you can ascertain the correct specification for cables when you start planning the installation.

Back to our holiday home example. Based on our previous calculations, our holiday home needs a solar array capable of generating 320 watt-hours of energy, if we angle the array towards the sun. This solar array will take up approximately $2m^2$ (18 sq. ft) of surface area.

Our site survey for the holiday home showed the main pitch of the roof is facing east to west. This is not ideal for a solar array. The eastern side of the roof has a chimney. There is shade from a tall tree that obscures part of the west-facing part of the roof. There is no space on the gable end of the roof to fit the required solar panels.

SITE SURVEY FOR THE HOLIDAY HOME

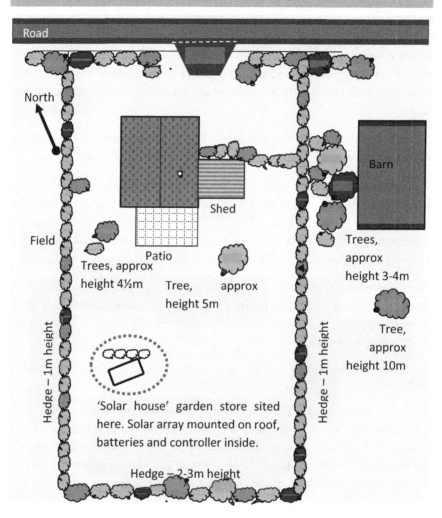

Map of the holiday home, identifying likely obstacles and a suitable position for the solar array

The garden is south-facing and receives sunlight throughout the year.

A shed close to the house has a south-facing roof. A tree shades the shed for most of the mornings for nine months of the year. Furthermore, the condition of the shed means that it would need remedial work should we decide to use it.

There is a farm to the east of the house, with a large barn and several trees bordering the house, the tallest of which is approximately 33 feet (10m) tall. One trees shades the shed and part of the rear of the house during the winter and may provide more shade during the rest of the year if it continues to grow.

We decide to install the solar array in the rear garden, constructing a suitable 1.2m (4 feet) tall garden store with a south-facing pitched roof of approximately 52° (allowing us to tilt the solar array at 38° from vertical for best year-round sunshine). Our *solar house* will hold the batteries and controller and will have adequate ventilation to ensure that the small amounts of hydrogen generated by the batteries can escape. By building our own structure, we can install the solar array at the optimum tilt to capture as much sunlight as possible. This means our solar array is compact and keeps our costs to a minimum.

The solar house will be located around 10m (33 feet) from the house and shielded from the house by a new shrubbery. The cable lengths between the solar array and the solar controller are approximately 2m (6½ feet). The cable length between the solar controller and the batteries is less than 1m (3 feet). The cable length between the solar house and the home is 12m (40 feet). There is a further 10m (33 feet) of cabling inside the house.

These longer cable lengths are not ideal. Cable runs should be as short as possible to reduce power losses through the cable. However, we cannot position the solar array any nearer to the house. We will have to address this problem through our design.

IN CONCLUSION

- There is a lot to do on a site survey. It is important. Spend time; get it right.
- Drawing a map and taking photographs can help with the site survey and are invaluable for the next stage, when we start designing our new system.
- Solar panels can be roof-mounted, or mounted with a frame or on a pole.
- Once you have identified a location for the panels, check for obstructions that will shade the panels throughout the year.
- These obstructions are most likely to be an issue during the winter months, when solar energy is at a premium.
- Identify a suitable space for batteries, controller and inverter.
- Plan the cable runs and the measure the length of the required cables.
- Cables should be as short as possible, to reduce the voltage losses through the system. If long cable lengths are necessitated by the positioning of the solar array, we may need to run our system at a higher voltage to compensate.

UNDERSTANDING THE COMPONENTS

Once you have completed your site survey, you know how much power you need to generate, the suitability of your site and approximately how much it is going to cost you. Now you need to look at the different technologies and products that are available, to see what best suits you and your application.

Your choice of components and the design of your system will depend on whether you are designing a stand-alone system (which also includes grid fallback and grid failover systems), or a grid-tie system that exports energy to the grid.

Because there are differences in the design of stand-alone and grid-tie systems, I have split this section into four chapters. This chapter looks at components that are common to both grid-tie systems and stand-alone systems. The next chapter looks specifically at what you require for grid-tie systems. The third chapter looks specifically at stand-alone systems and systems that incorporate their own battery store, whilst the fourth chapter looks at the component certification and regulations that you need to take into account when selecting solar energy equipment.

How to use these chapters

These next three chapters will go into much more detail about the different options available to you. There is an extensive choice of solar panels, batteries, controllers, inverters and cables.

These chapters explain the technology in a lot more detail, so you can go and talk sensibly to suppliers and understand what they are saying.

COMMON COMPONENTS FOR ALL SYSTEMS

Core to all solar energy systems are the solar panels themselves. Most solar panels can be used for either grid-tie or stand-alone use, and although recently some manufacturers have released higher-voltage solar panels designed specifically for grid-tie applications, the criteria for choosing one solar panel over another remain the same.

SOLAR PANELS

There are three different technologies used for producing solar panels. Each has its own set of benefits and disadvantages.

For the purpose of this handbook, I am ignoring the expensive solar cells used on satellites and in research laboratories and focusing on the photovoltaic panels that are available commercially at reasonable cost today.

Amorphous solar panels

The cheapest solar technology is amorphous solar panels, also known as *thin-film* solar panels.

These panels have had a bad reputation in the past, with poor product reliability and questionable lifespan. This has often been down to the chemistries used in older designs of panel breaking down under extremes of temperature over a period of a few years, or the poor quality of materials used in the production of cheap panels.

Thankfully, this technology has matured significantly over the past few years and amorphous solar is now regarded as being highly reliable, with some significant benefits over traditional solar panels. Big name manufacturers such as Sanyo and Sharp now manufacture high-quality amorphous solar panels, along with some exceptionally good specialist manufacturers such as Solar Frontier and Uni-Solar. Some manufacturers now even offer a ten or twenty-year warranty on their amorphous panels.

On paper, amorphous solar panels are the least-efficient panels available, typically converting around 6–8% of available sunlight to electricity. This means that you need twice as much space available for installing amorphous solar panels compared to crystalline panels.

However, amorphous panels are good at generating power even on overcast and extremely dull days. In general, they are also far better in extreme temperature conditions, with significantly less power loss at higher temperatures than other solar panel technologies.

Unlike other solar panel technologies, amorphous solar panels provide excellent performance even when partially shaded. Whilst a best-case scenario is to eliminate shading whenever and wherever possible, amorphous panels continue to operate at a high level of efficiency even if part of the array is in shade.

Amorphous panels can also be manufactured into a shape or mounted on a curved surface. They can be made to be hard-wearing enough to be fitted onto surfaces that can be walked on. A few solar manufacturers have started manufacturing amorphous solar roof tiles (or shingles), so that new-build houses can incorporate solar into the structure of the roof.

This combination makes amorphous panels suitable for integration into consumer products such as mobile phones and laptop computers, and for mobile products such as the roof of an RV or caravan, where the manufacturer has no control over where the products are placed or how they are used.

Amorphous panels are the cheapest panels to manufacture and several manufacturers are now screen-printing low-cost amorphous solar films. Between 2008 and 2013, amorphous solar panel costs dropped by around 30% each year. Whilst the prices are dropping more slowly now, the success of amorphous solar panels have been a big factor of driving down the costs of the entire solar industry.

Because of their lower efficiency, an amorphous solar panel has to be much larger than the equivalent polycrystalline solar panel. As a result, amorphous solar panels can only be used either where there is no size restriction on the solar array or where the overall power requirement is very low.

In terms of environmental impact, amorphous panels tend to have a much lower carbon footprint at point of production, compared to other solar panels. A typical carbon payback for an amorphous solar panel would be in the region of 12–30 months.

Most amorphous solar panels have comparatively low power outputs. These panels can work well for smaller installations of up to around 300-watt outputs, but not so well for larger installations where larger numbers of panels will be required and the additional expense in mounting and wiring these additional panels starts to outweigh their cost advantage.

Consequently, amorphous solar panels are often more suited to OEM applications, as an energy source built into a manufactured product, or for large-scale commercial installations where the panels are incorporated into the structure of a roof-space on a new build.

Some of the most exciting advances in solar technologies over the past three years have come from amorphous technology. Products as diverse as mobile phones, laptop computers, clothing and roofing materials have all had amorphous solar panels built into them. An exciting technology, amorphous solar is going to get better and better over the coming years.

Polycrystalline solar panels

Polycrystalline solar panels are made from multiple solar cells, each made from wafers of silicon crystals. They are far more efficient than amorphous solar panels in direct sunlight, with efficiency levels of 12–17%.

Consequently, polycrystalline solar panels are often around one third of the physical size of an equivalent amorphous panel, which can make them easier to fit in many installations.

Polycrystalline solar panels often come with a 25-year performance warranty and a life expectancy of far longer. Commercial solar panels only became available in the mid-1970s and many of these panels are still perfectly functional and in use to this day.

The manufacturing process for polycrystalline solar panels is complicated. As a result, polycrystalline solar panels are expensive to purchase, costing around 15% more than amorphous solar panels. The environmental impact of production is also higher than amorphous panels, with a typical carbon payback of 2–5 years.

Polycrystalline solar cells are dark blue and are typically produced with either a white or silver frame. This makes it more difficult to blend the panels into their surroundings, which can be an issue for some applications.

Prices for polycrystalline solar panels are dropping, thanks to both the increase in manufacturing capacity over the past few years and the increasing popularity for larger screen televisions, which use the same specification glass. Between 2007 and 2014, prices dropped by around 25-30% per year. In 2015 and 2016, prices dropped a further 10-15%. Prices have since stayed relatively static and are expected to remain at their current level for the next few years.

Monocrystalline solar panels

Monocrystalline solar panels are made from multiple smaller solar cells, each made from a single wafer of silicon crystal. These are amongst the most efficient solar panels available today, with efficiency levels of 15–20%.

Monocrystalline solar panels have the same characteristics as polycrystalline solar panels. Because of their efficiencies, they are the smallest solar panels (per watt) available. The cells themselves are a uniform black and are produced with either a black or silver frame. These have a smarter appearance than the polycrystalline panels and are very popular with homeowners.

Monocrystalline solar panels are the most expensive solar panels to manufacture and therefore to buy. However, the prices are still competitive, typically costing around 10% more than the equivalent polycrystalline solar panels.

Hybrid solar panels

Hybrid solar panels combine monocrystalline solar cells and amorphous thin-film *between* each monocrystalline cell. This provides the benefits of the efficiencies of monocrystalline and the benefits of improved shade and high temperature performance from the amorphous technology.

Hybrid panels provide efficiency levels of between 18—22% in optimum conditions, but really benefit from sub-optimal conditions where the real world performance can be 10-20% better than other solar panels.

The downside is price. Price-wise, hybrid solar panels are expensive, around twice the price of monocrystalline panels. Unless space is at such a premium that the price difference can be justified, hybrid solar panels are usually ruled out on economic grounds.

Solar Panel sizes and power outputs

Individual amorphous solar panels usually have a power output of between 5Wp and 100Wp. A 100Wp amorphous solar panel will measure somewhere around 150cm x 100cm (60 x 40 inches).

Polycrystalline and monochrystalline solar panels have higher power outputs. The smallest panels tend to have a power output of around 20Wp, whilst the largest panels will typically have a power output of around 350-400Wp.

The most commonly used crystalline panels tend to be the 250Wp panels, which measure approximately 160cm x 100cm (64 x 40 inches). The reason this size has become so popular is that it proved to be the most cost effective to manufacture: larger panels require more complex glass manufacturing capabilities, whilst smaller panels have little cost savings to the manufacturer. Consequently, the cheapest solar panels for most applications are 250Wp polycrystalline or monochrystalline panels, often costing around half as much as a 280-300Wp panel.

Which solar panel technology is best?

For most applications, monocrystalline panels offer the best solution, with reasonable value for money, compact size, good overall performance and smart appearance.

Amorphous panels can be a good choice for smaller installations where space is not an issue. They are usually not practical for generating more than 100-300 watts of power because of their overall size, unless you have an extremely large area that you can cover with solar panels.

What to look for when choosing a solar panel

Not all solar panels are created equal, and it is worth buying a quality branded product over an unbranded one. Cheaper, unbranded solar panels may not live up to your expectations, especially when collecting energy on cloudy days.

If you are spending a lot of money buying a solar energy system that needs to last many years, it is advisable to purchase from a known brand such as Eco Future, Kyocera, Panasonic, Hyundai, Mitsubishi, Solar Frontier or Sharp. My personal recommendation is Eco Future monocrystalline solar panels, or Solar Frontier amorphous panels. I have found these to be particularly good.

Buying cheap solar panels

Not all solar energy systems have to last ten or twenty years. If you are looking for a small, cheap system to provide power to an RV or caravan, or your requirements are modest, such as installing a light in a shed, buying a cheap solar panel may well be the right option for you.

The quality of the cheaper solar panels has improved significantly over the past few years. Six or seven years ago, buying a cheap, unbranded Chinese-made product was a recipe for disaster. Many of the panels were poorly assembled, allowing water to seep through the frames and damaging the solar cells. A lot of them used plate glass, often a thin, low-grade glass that becomes clouded over time and is easily chipped or broken. The cells used by these manufacturers were often sub-standard reject cells and often degraded very quickly.

Thankfully, most of these problems are now resolved and if you buy a cheap solar panel on eBay, you are likely to have a good product that will reliably generate power for five to ten years, and in all probability a lot longer. If you are buying a solar panel from a manufacturer you have never heard of, here is a checklist of things to look for:

Buy bigger than you think you'll need

If you are buying a very cheap solar panel, you are likely to be saving as much as 50% of the price when compared to buying a branded product. However, expect it to degrade slightly more quickly than a branded unit, and do not expect it to be quite so efficient.

To counter this, buy a solar panel with a higher watt rating (often shown as a watt peak, or Wp, rating) than you need, or buy additional solar panels if you are purchasing an entire array. Aim for 15% more power than you would otherwise have bought. You will still save a lot of money, but you will have an extra bit of assurance that the system will be up to the task.

Warranty

With cheap solar panels, you're not going to get a five-, ten- or twenty-year warranty, but you should still expect a one- or two-year warranty with any solar panel you buy. Check to see exactly what the warranty offers.

You are looking for a warranty that guarantees a minimum output under controlled conditions. The standard across the industry is to guarantee 80% of the quoted output under controlled conditions.

If you have a warranty claim, also check to see how you can claim on that warranty. Shipping a broken solar panel half way around the world and paying for return carriage is likely to cost as much as buying a new solar panel.

Glass

Some cheap solar panels are cheap because they have skimped on the glass. Always make sure that the solar panels you are buying use tempered glass.

Tempered, or toughened, glass is around eight times stronger than plate glass. This makes it far more robust. If your glass is chipped on your solar panel, you will immediately see a significant drop in power output. If water gets into the solar panel itself, it can create a short circuit and becomes a fire hazard. Water and electricity do not mix.

It is also worth investing in solar panels that use self-cleaning glass. Self-cleaning glass has a low-friction coating that ensures the panels are washed every time it rains. This ensures your panels remain clean and therefore more efficient in every day conditions.

Second-hand solar PV panels

From time to time, second-hand solar panels appear for sale. They appear on eBay or are sold by solar equipment suppliers or building salvage yards.

So long as they come from a reputable brand, second-hand solar panels can be extremely good value for money and even old panels that are 25–30 years old may still give many more years of useful service. Although good quality solar panels should provide at least 25 years service, nobody knows how much longer they will last. The early commercially available solar panels (which are now over forty years old) are still working extremely well, typically working at around 70% of their original capacity.

There are, however, a few points to look out for if you are considering buying second-hand solar PV panels:

- Never buy second-hand solar PV panels unseen. Take a multi-meter with you and test them outside to make sure you are getting a reasonable voltage and wattage reading
- Check the panels and reject any with chipped or broken glass. Also reject any panels where the solar cells themselves are peeling away from the glass or have condensation between the glass and the solar cell
- The efficiency of older solar PV panels is significantly lower than new panels. 30 years ago, the most efficient solar panels were only around 5–6% efficient, compared to 13–24% efficiency levels today. 10–15 years ago, the best solar panels were around 10–12% efficient. Consequently, a solar PV panel from the early 1980s is likely to be three times the size and weight of an equivalent modern crystalline panel
- These second-hand panels will not have any of the safety certification ratings that you get with new solar panels. This may cause issues with building regulations or building insurance, if you are installing these onto a building as part of a new solar installation.

Concentrated Photovoltaics and Solar Concentrators

Concentrated Photovoltaics (CPV) technology uses Fresnel lenses and curved mirrors to concentrate a large area of sunlight onto a smaller area of photovoltaic cells. They do this by refracting the light to concentrate it and increase the intensity of sunlight in a smaller area.

In effect, by concentrating the sunlight into a smaller area and increasing the solar irradiance, significantly more energy can be captured by the solar panel, thereby improving its efficiency quite impressively.

However, there are problems with this technology. Most specifically, the heat build-up is quite considerable and, in testing, many solar panels have been destroyed by the excessive heat generated by the concentrated sunlight. This is especially true of Fresnel lenses built by enthusiastic amateurs.

The panels tend to be quite large and bulky. Due to the heat build-up, they also need to be very carefully mounted, with adequate ventilation around the panel.

There are now a few solar concentrators on the market, but the technology has, by and large, being overtaken by events. The logic behind their design was that solar panels were expensive and lenses and mirrors were cheap. As the price of solar has dropped so much over the past few years, the economics for solar concentrators no longer add up.

A few installations use mirrors or polished metal to reflect additional sunlight back onto solar panels and therefore increasing the solar irradiance. However, you must take care to ensure that the reflected light does not dazzle anyone.

SOLAR PANEL MOUNTINGS

You can choose to fabricate your own mountings for your solar panels if you are planning a stand-alone system, but it is far more common to purchase the mountings at the same time as purchasing your solar panels. The main options for solar panel mountings are roof mounting, frame mounting or pole mounting.

How solar panels are mounted to a frame

Some smaller solar panels – typically up to 40Wp capacity – often incorporate plastic frames with holes for bolting the panels to a frame. Most solar panels, however, are designed to be clamped to a frame. The frame is typically made up of two mounting rails. Clamps are used to mount the solar panels to the rails. The rails in turn are either mounted onto a roof structure using roof hooks, or form part of a larger frame structure for ground mounting or for mounting on a pole.

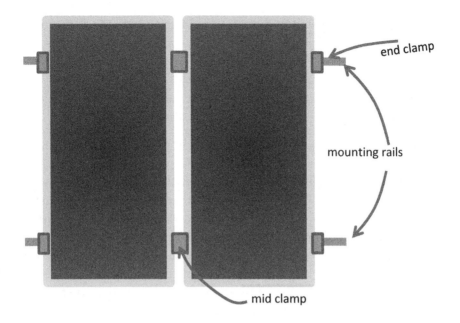

Self-locking nuts should be used to ensure the panels remain firmly clamped. Solar panels suffer from wind buffeting over time, which can loosen bolts. Some frame clamps are designed to vibrate tighter with buffeting, reducing the risk of panel damage over time.

Roof Mounting

In a roof mounted system, the mounting rails are mounted onto the roof using roof hooks. Roof hooks mount directly onto the roof trusses underneath the roof tiles. To install them, individual tiles are slid out from the roof and

Below: Close up of a roof-mounted system, showing a roof hook, the mounting rail and a solar panel clamp.

the roof hooks are mounted to the roof truss underneath. The tiles are then slid back into position.

Frame Mounting

For ground mounting, or for mounting onto a flat roof, solar panels are typically installed onto a freestanding frame, or on a moulded plastic tray.

A freestanding frame can either fit just one solar panel, or multiple panels together. The benefit of freestanding frames is that the angle of the solar panels can be adjusted for the location to make the best use of the sunlight. Some freestanding frames allow the angle of the panels to be adjusted throughout the year to get the best performance from the available sunlight. Freestanding frames need to be anchored to whatever they are mounted onto. If mounted on the ground, frames are typically mounted onto concrete foundations, whereas if they are mounted onto a flat roof, mounting bolts are often required to penetrate the roof. The mountings have to be fairly substantial to counter wind buffeting that can lift the panels if not securely bolted down.

Moulded plastic trays are often much easier to install than freestanding frames. The solar panels are mounted onto the trays and the trays are weighed down with ballast – usually in the form of bags of sand. This has the benefit that you do not need to mount the solar panel frames to a building, speeding up the installation process. The trays are moulded for particular angles and you buy the plastic tray that best suits your application.

Left: A simple mounting frame for a single solar panel.
Right: The PV-Pod uses water as a ballast, simplifying installation.

Pole Mounting

Pole mounting your solar array can be a practical option and can often be cheaper than roof mounting. Some pole mounted systems are designed to mount to existing poles, such as lamp-posts, and suitable for one or two solar panels, other systems are designed to be dedicated solar mounting systems and can mount multiple solar panels.

Some freestanding frames, such as these from Crown International, allow you to either manually or automatically adjust the solar panel tilt across the year.

Some pole mounted systems are designed to be easily adjusted, so that the angle of the panels can be adjusted throughout the year to get the maximum power generation from the available sunlight. This can significantly increase the amount of power generation from your solar array over a period of a year.

The size and thickness of the pole depends on the number of panels required and the height of the pole. The pole must be anchored into the ground, set into concrete. The depth of the foundation must be at least one half of the height of pole above the ground. For example, if your solar pole mounting is going to be 2 metres high (around 7 feet), the depth of the hole has to be a minimum of 1 metre (3½ feet).

Solar Trackers

Solar panels generate the most energy when they are facing directly at the sun. Solar trackers are mounted on a pole mounted system to move the panels throughout the day to capture the maximum amount of sunlight.

Some solar trackers simply move the angle of the panels from left to right (single-axis trackers), whilst more advanced trackers adjust both the angle from left to right and the tilt (double-axis trackers).

Whilst solar tracker manufacturers can claim a performance benefit of between 35-50% for their products, most installers report a performance benefit of around 30% ($^+/_-$5%) for single-axis trackers, and around 40% ($^+/_-$5%) for double-axis trackers. The performance improvements are best during the summer months, when energy capture can be increased by as much as 55%, whilst during the winter months, energy capture is improved by around 15–20%.

Solar trackers work using sensors and electric motors or hydraulics, consuming a small amount of electricity. The amount of energy they consume is a small fraction of the additional energy they capture from the sun's rays.

Consequently, adding solar tracking for a solar array can have significant real world performance improvements. This explains their popularity with large solar farms. However, for smaller installations, their benefits need to be weighed against their costs. Unless space is a premium, buying many more solar panels can work out much cheaper than buying a smaller array and installing solar tracking.

Mounting solar on vehicles

If you want to install solar panels onto a vehicle, such as a caravan/travel trailer or recreational vehicle, you need to take airflow into account. You need to reduce the drag of the solar panels on the roof of your vehicle, whilst ensuring there is adequate airflow around the solar panels to reduce the risk of heat build-up. If you do not take airflow into account, you may have additional wind noise when your vehicle is in motion, or at worst run the risk of having the solar panels ripped off the roof by the wind.

There are two main options for fitting solar onto a vehicle. The first is to bond flexible solar panels directly onto the roof of the vehicle. This works best if bonding onto a metal surface rather than glass fibre or plastic as the metal surface acts as a heat-sink allowing the solar panel to dissipate heat.

The second option is to mount your solar panels with a spoiler mounting. These mounts deflect the air over the solar panel, leaving an air gap beneath the panel to ensure the panel does not overheat.

These spoilers are often panel specific, so make sure your supplier can provide them for the solar panels you are planning to buy. These mountings are typically bonded onto the roof of the vehicle, using a high strength glue such as Sikaflex 252. Once the glue is dry, this creates an extremely strong and permanent mounting for the panels that will be easily strong enough to withstand the high speed wind buffeting from a moving vehicle.

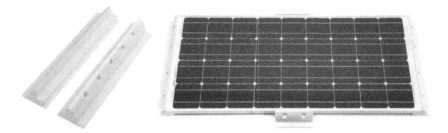

Left: Vehicle spoiler mounting.
Right: A single 80Wp solar panel mounted with spoiler mountings.

Single panel installations for outbuildings

If you are planning to fit one or two panels to an outbuilding, such as a garden shed, you can buy simple plastic corner mountings that can either be bonded or screwed into place. These mountings are similar to vehicle spoiler mountings and are used in the same way.

Fabricating your own mountings

Some people choose to fabricate their own mounting for your solar panels. I have done this occasionally, although I now usually purchase ready-made mountings. Incidentally, if you are fitting solar panels to a building, you must always use a professional mounting system.

If you are designing your own mounting system, the design must take into account wind loading, so that it is not damaged or destroyed in high winds. Mounting systems must also ensure there is adequate ventilation behind the panel to avoid excessive heat build-up. If you are using smaller solar panels – up to around 100Wp panels – an air gap of around 2 – 2½cm ($^3/_4$ – 1 inch) is sufficient. For larger panels, an air gap of 5 – 8cm (2 – 3 inches) is usually recommended.

You will need to make sure that your mounting system is either strongly anchored to the ground, or heavy enough that it does not move in high winds. Due to the design of a solar panel, wind that blows underneath the panel through the air gap will travel faster than wind that blows across the top of the panel. This creates an area of low pressure underneath the panels, creating a lift effect, similar to that of an aircraft wing. Poorly secured solar panels can be lifted off their mountings, or pull their mountings with them, in high winds.

Consequently, you need to ensure that the mountings are capable of withstanding a lifting force equal to twice the weight of the solar panels they are holding.

SOLAR ARRAY CABLES

Solar array cables connect your solar panels together and connect your solar array to the solar controller. These cables are often referred to as 'array interconnects'. You can purchase them already made up to specified lengths or make them up yourself. The cables are extremely heavy duty and resistant to high temperatures and ultra-violet light. They also have a tough, extra-thick insulation to make them less prone to animal damage.

If you are planning to wire your solar array in parallel rather than in series, you need to ensure that your solar array cables can cope with the current that you are going to be generating through your solar array. If you are designing a parallel design system, I explain how you can calculate the size of cable required in the chapter on stand-alone system components. You can read about this on page 128.

SOLAR ARRAY CONNECTORS

There are standard connectors for connecting solar panels together. For most solar panels, the standard connector is the MC4 connector, as shown in the pictures below. The MC4 connector comes as male and female components that clip firmly together, creating a weathertight seal.

For very small solar panels, such as the 5Wp – 60Wp panels used for outbuildings and the like, the alternative S-S connectors are common. These are cheap and simple connections that are suitable for 12V systems only with a maximum 5A current, and are typically used where only one or two small solar panels are being used, with shorter cable lengths of up to 3 metres (10 feet).

Like the MC4 connections, S-S connectors form a weathertight seal. Unlike the MC4 connections, they do not lock together and can be easily pulled apart. They are suitable for very small installations and are easy to use.

FUSES AND ISOLATION SWITCHES

The ability to isolate parts of the system is important, both for installation and maintenance. Even comparatively low voltages can be dangerous to work on.

Even small stand-alone systems should incorporate fuses and switches at critical paths in the system. This allows the system to shut down automatically if there is an overload or a short circuit, or be manually switched off if there is a problem. It is far better to blow a cheap fuse than to fry a battery or an expensive solar controller. At the very least, you should have:

- An isolation switch between the solar array and the solar controller or solar inverter, so that the power generation from the solar panel can be shut off.
- A quick-blow fuse between the battery pack and the solar controller.
- An isolation switch and a quick-blow fuse between the battery pack and anything that is taking power from the batteries (such as an inverter).
- If you are using an inverter, an isolation switch must also be fitted on the AC side of the system.

Even if you are only installing a very small solar energy system, to provide a shed light for example, adding these fuses and switches is important. For systems running at 12V that can only generate or use less than 100W of power, the switches can be low cost home light switches and automotive blade fuses. For anything larger, specialist DC isolation switches and fuses will be required.

If your solar panels are mounted some way from your inverter or controller, it may also be a good idea to have an isolation switch fitted next to the solar panels, as well as one fitted next to the inverter or controller. You can then easily disconnect the solar panels from the rest of the system for maintenance or in case of an emergency.

Ensure that the isolation switch you choose is capable of handling high-current, high voltage DC circuits, with contacts that will not arc. AC isolation switches are not suitable and should be avoided at all costs. Suitable DC isolation switches are available from any solar supplier and are not expensive to buy.

If you are planning a grid-connected system, you will need AC isolation switches to allow you to disconnect the inverter from the grid supply. You will require an isolation switch next to the inverter, and a second one next to the distribution panel.

THE IMPORTANCE OF USING THE CORRECT CABLES, CONNECTORS AND ISOLATION SWITCHES

Solar arrays are often very high voltage. Even a small off-grid installation can often have voltages of 100–150V, whilst a larger residential installation can easily have a voltage as high as 500–900V on a bright day.

Unlike electricity from the grid, that uses an alternating current, solar electricity is a direct current. If you get an electric shock from an alternating current, the likelihood is that the force of the jolt will throw you back, thereby ensuring that you only receive a momentary shock. With a direct current electric shock, you will not be thrown back. Instead, your muscles are much more likely to lock you into the same position. You are therefore much more likely to be electrocuted for longer, which poses a much greater risk.

Because of the very high voltages, the electricity can arc across quite wide gaps between cables. I have seen an arc of 10–15cm (4–6 inches) when somebody has disconnected solar array connectors without first breaking the circuit through an isolation switch.

For the same reasons, you must use isolation switches that are rated for high voltage DC connections. Usual residential isolation switches are designed for AC power, which will not arc, and are entirely unsuitable.

The cables and connectors have a harsh life. They are usually outside and are expected to withstand harsh weather, intense UV light, heat, cold and potential attack from rodents and small animals, and they are expected to last for between 30 and 40 years. Do not be tempted to use non-standard cables or connectors for your solar installation, no matter how small your system is. The cost of getting the right equipment in terms of cables, connectors and switches is low, and the potential risks of using unsuitable cables, connectors or isolation switches is not worth taking.

GROUND FAULT PROTECTION

Ground fault protection ensures that if there is a short within the solar array, the current flow is cut off immediately. This averts the risk of damage to either the controller or the solar array, and significantly reduces the risk of electrocution.

Ground fault protection works by measuring the current entering and exiting a circuit. If everything is working correctly, the current in should equal the current out. However, if there is a 'leak' or a partial short circuit, the system will see a difference in current and immediately shut down. A partial short circuit could occur if a solar panel was broken or if somebody touched an exposed cable.

Most solar inverters and solar controllers incorporate ground fault protection, using a *Residual Current Device* (RCD) built into the unit (note: RCDs are known as *Ground Fault Interrupters* – GFIs – in the United States and Canada). Many experts say that it is prudent to install a separate ground fault protector, even if the controller or inverter has ground fault protection built in. As the cost of an RCD or GFI is low and the benefits they provide are high, this is good advice.

You will require separate ground fault protection for your DC and AC circuits:

- For anything larger than 100Wp solar panel systems, and for all systems mounted to a building, you should install ground fault protection between your solar panels and your controller or inverter
- If you are installing a DC power supply into a building for running appliances, you must install ground fault protection between your controller and this power supply
- If you are using an inverter, you should install ground fault protection between your inverter and any load

There are specific RCD units for DC circuits and these are stocked by solar panel suppliers.

COMPONENTS FOR GRID-TIE SYSTEMS

Before discussing the components required for a grid-tie system in more detail, it is useful to look at how a grid-tie system is configured.

There are three basic designs for grid-tie solar energy systems:

- Single string in-series systems
- Multiple string in-series systems
- Micro-inverter systems

Single string in-series systems

Multiple solar panels connected in series is called a 'string' of panels. For grid-tie systems, a single string of solar panels is the most popular configuration for a grid-tie system today. Solar panels are connected together in series, producing a high-voltage DC power. This is then fed into a central inverter to convert the power into an AC source, which in turn is connected into the standard building electrical system:

Solar panels connected in series (string)

Grid-tie inverter

AC power

A simplified block diagram showing the basic layout of a high voltage in-series solar energy system

This design is the most cost-effective design for grid-tie systems. It is relatively straightforward to install, simple to maintain, and components are readily available. By running the solar array at high voltage, it is also very efficient, with minimal losses through the array itself and allowing the inverter to run at a very high level of efficiency. This is why this design is currently so popular within the grid-tie solar industry.

This high-voltage DC power has the benefit of great efficiency, but comes with a number of very significant safety risks. Voltages as high as 600 volts in North

America and 1,000 volts in Europe are common. These voltages can very easily be fatal on contact.

These high voltages can also cause significant problems if there is damage to the wiring between solar panels, either due to a mistake during installation, through animal damage, or simply through wear over time. If a damaged cable generates a high-voltage, direct current electric arc whilst the panels are in direct sunlight, the immensely high temperatures can easily melt metal and are a potential fire hazard.

When connected in series, the solar array is only as strong as its weakest link. This is a significant disadvantage of in-series connection. If you have a damaged solar panel, shade blocking light to a few solar cells, or a damaged cable, the output of the entire solar array drops to the output of that weakest link.

Multiple string in-series systems

It is for these reasons that multiple string in-series systems are becoming more popular. Multi-string inverters allow you to connect more than one string of solar panels. These two strings then work as separate arrays, but feed the power through the same inverter.

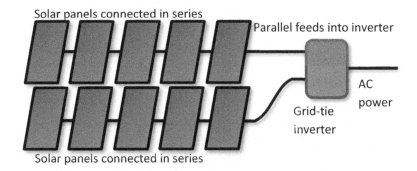

A simplified block diagram showing the basic layout of a low-voltage solar energy system where multiple strings of solar panels are connected in parallel

Because your two separate strings work as independent solar arrays, you can have two different sizes of array with different solar panels, or have the arrays mounted on different roofs or at different orientations. If one array is partially shaded, the performance does not impact the second array.

This design used to be more expensive than a single string of panels because it required a more expensive grid-tie inverter that could accept multiple strings of solar panels. However, over the past two years, the price of these inverters has dropped significantly and there is now very little difference in cost between a grid-tie inverter that can only accept one string and grid-tie inverters that can accept two strings of solar panels.

This approach is cost effective and gaining in popularity over the single-string systems. Multiple string systems are particularly beneficial in the following situations:

- Where you wish to fit two different sizes of solar panel
- Fitting solar panels facing at different orientations, such as on a roof that has two different pitches
- Resolving shading issues. Solar panels that are in shade at certain times of the day can be put onto a separate string so as not to affect the output of the rest of the system

Micro-inverter systems

Micro-inverter systems have been around for a few years now, but until very recently had not gained popularity due to their higher cost.

However, with the huge growth in the popularity of grid-tie systems around the world, micro-inverter systems have become far more widespread as the benefits of this technology have become more apparent. Prices for micro-inverter systems are dropping fast, although they are still more expensive than single inverter systems.

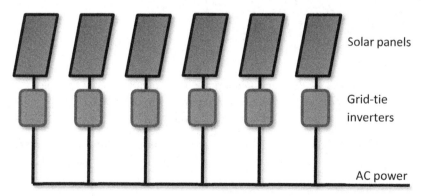

A simplified block diagram showing the basic layout of a micro-inverter system

In a micro-inverter system, each solar panel has its own inverter, converting its output to AC power. In effect, each solar panel becomes its own independent solar energy system.

In most micro-inverter systems, the inverter itself is mounted outside, bolted to the frame that also holds the solar panel in place. The individual solar panels are connected to an AC power cable that runs alongside the solar array and then feeds directly into the building's electrical system.

There are some very significant benefits of micro-inverter systems over other forms of grid-tie systems:

- As each solar panel runs as an independent unit, if one panel is under-performing, either because of damage or simply because it is shaded, it does not affect the output from any of the other panels
- Because there is no high-voltage DC power running from one panel to the next, safety is less of an issue
- Installation, fault-finding and maintenance also become significantly easier
- It is easy to expand your grid-tie system in the future, as budget allows
- More flexible solar panel installation – you can have solar panels mounted in different locations and facing in different directions

Whilst today, series-connected solar systems with a central inverter are still the standard model for installing grid-tie systems, the industry is moving towards micro-inverter technology as a far better model.

Now you have a clearer idea of how a grid-tie system is put together, it is time to look at the components available in more detail.

GRID-TIE SOLAR PANELS

In the past, when most solar energy systems were stand-alone systems, almost every solar panel you could buy was rated for a 12-volt output. Whilst this is still true for smaller panels, there are now higher-voltage configurations available for larger solar panels.

As grid-tie systems have become more popular, higher-voltage solar panels have become available. Many solar panels of 150Wp capacity and over are rated for a 24-volt output and some manufacturers are now building solar panels with rated outputs of between 48 volts and 120 volts.

These higher voltages are well suited to grid-tie installations. By running your solar array at a higher voltage, you can keep the current flow low, which improves the efficiency of the overall system.

GRID-TIE INVERTERS

Grid-tie inverters convert the DC power from your solar energy system into AC power, and convert the voltage to the same as the grid. This allows you to connect your system into the grid, enabling you to become a mini power station and supply your electricity to the electricity companies.

You cannot use an ordinary inverter for grid-tie applications. There are a number of reasons for this:

- Grid-tie inverters have to work in conjunction with the grid, to be able to export electricity to it. The AC pure sine waveform generated by the inverter has to be perfectly coordinated with the waveform from the grid
- There is an additional safety feature with grid-tie inverters to cut off power from the solar array if the grid shuts down
- Grid-tie inverters are connected directly to the solar panels. In an in-series system, this means the input voltage from the panels can fluctuate wildly, often jumping or dropping by several hundred volts in an instant. Non grid-tie inverters cannot cope with such massive voltage jumps
- In many countries, grid-tie inverters have to be certified for use with the grid

There are a number of things to consider when purchasing a grid-tie inverter:

- Input voltage
- Power rating
- Power tracking
- How many strings the inverter can support
- Diagnostics and reporting information
- Inbuilt safety systems
- Installation options and operating environment
- Certification and local regulations

Input voltage

Your choice of inverter will have a large voltage range to cope with the huge fluctuation of voltage that a solar array can provide. From this voltage range, you will be able to identify how many solar panels the inverter can cope with, when connected in series.

Most grid-tie inverters are designed to work with a wide range of input voltages. A range of between 50 volts and 500 volts per string is common for a residential system.

Remember that the rated voltage of a solar panel is not the maximum voltage that the solar panel can generate. The voltage from a single 12-volt solar panel can fluctuate anywhere from 12 volts on a very dull day, up to around 20 volts in intense overhead sunlight. If you have a 48-volt solar panel, or four 12-volt solar panels connected together in series, the voltage swing can be between 48 volts and 88 volts.

In addition to this, a solar panel can produce significantly higher voltages in an open circuit – i.e. when the solar array is generating power but the power is not being used. Depending on your solar panel, it is possible for a single 12-volt solar panel to generate 26 volts in an open circuit.

As you can see from the table below, the higher the nominal voltage from your solar array, the greater the voltage fluctuation can be:

Number of 12-volt solar panels	Nominal solar array voltage	Low voltage on dull day	Peak voltage in intense sunlight	Maximum open-circuit voltage
1	12-volt	12 volts	20 volts	26 volts
2	24-volt	24 volts	40 volts	52 volts
4	48-volt	48 volts	80 volts	104 volts
6	72-volt	72 volts	120 volts	156 volts
8	96-volt	96 volts	160 volts	208 volts
10	120-volt	120 volts	200 volts	260 volts
15	180-volt	180 volts	300 volts	390 volts
20	240-volt	240 volts	400 volts	520 volts
25	300-volt	300 volts	500 volts	650 volts
30	360-volt	360 volts	600 volts	780 volts
35	420-volts	420 volts	700 volts	910 volts
40	480-volts	480 volts	800 volts	1,040 volts

Note: the maximum open-circuit voltage allowable in the United States for grid-tied systems is 600 volts, whilst in Europe it is advisable that your system does not exceed 1,000 volts. You must ensure that your system never exceeds this.

As you can see from this table, if a heavy cloud blocks the sun on an otherwise clear day, you can see a voltage drop of several hundred volts in an instant. When the cloud passes over, the voltage shoots back up again.

It is important to ensure that the solar panel will work with the peak voltage of your solar array and not just the nominal voltage of your array. If you exceed the peak voltage of your inverter, the inverter will shut down to avoid damage. In extreme cases, you could damage or destroy your inverter by exceeding the input voltage rating.

As well as looking at the maximum input voltage, you need to look at the minimum input voltage as well. If the overall voltage of the system drops below the minimum input voltage, the inverter switches off. The solar panels still generate energy, but the energy generated is lost. A low minimum input voltage will mean more power generation at the start and end of each day and will ensure greater power generation in winter months, particularly in overcast conditions. Finding an inverter with the widest *range* of input voltages will give you the greatest flexibility in designing your system.

In addition to the standard input voltage range, your inverter will also show a maximum voltage rating. This maximum voltage rating relates to the maximum *open circuit* voltage of your solar array. You must ensure that the open circuit voltage of your array does not exceed the maximum voltage of your inverter.

Power rating

There are two power ratings on a grid-tie inverter:

- Input power rating – the minimum and maximum amount of power the inverter can accept from the solar array
- Output power rating – the maximum amount of power and current the inverter can generate as an AC output

Input power rating

The input power rating shows the minimum and maximum wattage range the inverter can work with. In the main, the wider the range, the more efficient your inverter is.

The specification on an inverter will typically show three figures for input power rating:

- A nominal power rating, shown in watts
- A minimum and maximum power range
- A start-up power rating

The nominal power rating shows the maximum amount of power that the inverter can convert into an AC output. If you exceed this figure on your solar array, the additional power will be lost and converted into heat. Exceed this figure for long and your inverter may shut down to avoid overheating.

The minimum power rating shows the minimum amount of power that your solar array must generate for the inverter to start producing power. The maximum power rating shows the maximum amount of power that can be fed into the inverter before you risk damaging your inverter.

The start-up power rating is the minimum amount of power the solar inverter requires to power itself. If the solar array produces less than this amount of power, the inverter will switch off.

Output power rating

The output power rating is the maximum continuous AC power that the inverter can generate. The output power information will show voltage, nominal output power in watts, the maximum output current in amps and the alternating current frequency.

For North America, the grid voltage is nominally set at 110 volts, with a frequency of 60Hz. For the majority of the rest of the world, the grid voltage is nominally set at 230 volts, with a frequency of 50Hz. In both cases, it is normal to get some variation in voltage and frequency.

The output power rating will also show the maximum efficiency rating of the inverter, given as a percentage. This rating is usually in the region of 90–94% with modern grid-tie inverters. If you are shopping on a budget, you must check this rating as some very cheap grid-tie inverters may be much less efficient.

It is not uncommon for a solar array to have a peak wattage higher than the output power rating from the inverter. In the United Kingdom for example, most companies selling residential solar energy systems will offer a 4kWp system as their largest system. However, the electricity companies have said that any system providing more than 16 amps must go through prior approval from the

District Network Operator (DNO), which typically takes up to eight weeks and can cost in the region of £500. 16 amps at a nominal 230V equates to 3.6kW. Most 4.0kWp residential systems installed in the United Kingdom therefore fit an inverter with a 3.6kW maximum power rating 3.6kW to comply with this directive. Due to the climate in the United Kingdom, a solar array with a 4kWp rating is unlikely to generate more than 3.6kW at any one time. In real terms, the customer is not losing out in real world power generation.

Power tracking

As discussed on page 67, the efficiency of the solar array depends on how efficiently the fluctuating voltage is handled by your inverter.

If purchasing a grid-tie inverter, you should invest in one that incorporates maximum power point tracking (MPPT). Maximum power point tracking can provide an additional 15–20% of energy compared to non-MPPT inverters.

Today, MPPT is the norm, but there are a few older designs of inverter that occasionally pop up for sale at bargain prices online. No matter how cheap these are, the performance loss rarely makes them a worthwhile investment.

Multiple strings

Multi-string systems were discussed on page 117. Many inverters now allow two strings of solar panels to be connected to the same inverter. Some allow 3-4 strings to be connected, although these tend to be significantly more expensive.

Diagnostics and reporting information

Almost all inverters provide a level of diagnostics and reporting information, either using a small display built into the unit itself, a series of LEDs on the front panel or a separate plug-in monitoring unit.

Many inverters now have a built-in internet connection, allowing them to connect to a wireless network. This means your system can provide you with updates via e-mail, via a built-in website, or even send updates to your mobile phone. In many cases, these systems can be remotely monitored by the solar inverter supplier, who can then notify you if there are any potential issues with your system.

If you have a micro-inverter system, the diagnostics and reporting system is usually a separate box, which is either connected to the AC power at the point where the solar array connects to your distribution panel, or it communicates wirelessly with the micro-inverters. This ensures that you have one central information point for your inverters.

As a bare minimum, you want an inverter that can tell you if your system is working and provide you with an indication of what may be the problem if a fault is detected. The diagnostics should be able to give you enough information to enable to you identify a fault with your system, such as:

This is one of my own test solar energy systems, being monitored online. The website allows me to see how much power is being generated at any one time as well as the overall energy generation of my system on a daily, weekly or monthly basis.

- Insufficient or excess power from the solar array
- Grid connection issues
- Grid voltage or frequency issues
- Overheating

Most solar inverters will provide you with much more information, allowing you to see the voltage and current from your solar array and the amount of AC power being generated at that moment. They will also be able to show the amount of energy generated by the system, both for that day and since the system was installed.

Built-in safety

Most inverters incorporate safety shutdown. It is common for inverters to have ground fault protection built in. As mentioned in the previous chapter, even if your inverter does provide this, it is still good practice to install additional ground fault protection when you design your system. This is incorporated into your system using a *Residual Current Device* (RCD). RCDs are known as *Ground Fault Interrupters* (GFIs) in the United States and Canada.

A grid-tie inverter will also monitor the power from the grid and shut down if it detects a power cut (sometimes referred to as 'Island Protection').

This power shutdown ensures that your solar system does not continue to feed power into the grid if there is a power cut. This is a necessary safety requirement. If workers are attempting to repair a power outage, they risk electrocution if power is being fed into the grid from your solar array while they are working.

Inverters should also shut down or derate if the internal temperature gets too high, to avoid permanent damage.

Installation options and operating environment

Inverters tend to be heavy units. They need to be mounted securely on a wall or bolted to the floor. They can generate a significant amount of heat, especially when running close to their rated output, and require good airflow around the units to keep them cool.

You can purchase inverters for either indoor or outdoor installation. If you are looking at installing outdoors, check that the unit is sealed against dust and water ingress, rated to at least IP64.

Overheating inverters is the number one reason for grid-tie systems failing. As an inverter gets hotter, it provides less power, and if the temperature continues to rise, it will eventually shut down to avoid permanent damage. When choosing an inverter, check its operating temperatures and consider how you can ensure your system remains within these limits.

Inverters are often installed in attics. Whilst this can be a very good idea, attics can get extremely hot in summer months. Even in the United Kingdom where hot days are something of a rarity, attic temperatures can reach 50–55°C (122–131°F) in some instances. Most inverters are getting to their upper limit of operating temperature at this sort of level.

Inverters should always be installed in a well-ventilated area, away from the ceiling and with a clearance around each side, the top and the bottom. They cannot be installed in a sealed cupboard. Some inverters have the option of an external heat sink or temperature-controlled cooling fans to help keep the inverter cool.

Some inverters do make a small amount of noise. This is typically a continuous low-level hum. It is usually only noticeable if the surrounding area is quiet. However, for this reason, inverters are not usually installed inside the living space in a home or in an office environment. Instead, consider installing your inverter in a garage or on an outside wall of your building.

Occasionally, the sound made by the inverter has been known to resonate with the wall, amplifying the sound and making it quite unpleasant to live with. This can occur even if the inverter is mounted to an outside wall. This is a very rare occurrence, but is most likely to occur if you are planning to mount your inverter onto a wall made of solid concrete. The solution is to dampen the mounting between the inverter and the wall or floor that the inverter is mounted on. There is a very effective product called *Green Glue*, produced by the Green Glue Company (*www.GreenGlueCompany.com*) that is applied between the wall and the inverter. When it is compressed by remounting the inverter, the glue spreads out to form a sound isolation barrier that is particularly effective at blocking out low-resonance vibrations.

Buying from eBay

Some companies have been selling non-approved grid-tie inverters online, most commonly on eBay. These are often sold at a bargain price, bundled with a cheap solar panel and often advertised as a 'micro grid-tie system'.

The sellers claim that these systems are designed for amateur installation. The inverter plugs into the household electricity supply through a normal domestic power socket, and the systems look exceptionally easy to install and use. The sellers often claim that you can use these systems to sell power back to the utility companies and that they can be used to run the meter backwards.

These systems are highly dangerous and must be avoided. The equipment has inevitably not been certified for grid-tie use in any country. More importantly, the use of these systems is illegal in the United States, Canada, Australia and across the European Community, because of the way the inverters connect to the household electricity supply, using a domestic power plug in reverse.

This means that the household plug has grid-level AC power running through it, with live high voltage current running through the prongs on the plug. This is extremely high risk and directly contravenes basic electrical safety legislation. In the United Kingdom, for instance, this is directly in contravention with BS7671:2008 (amd 1, 2011) 551.7.2 (ii). The catastrophic and potentially fatal results should somebody unplug the cable and accidentally touch the unshielded plug do not bear thinking about.

These cables have been nicknamed 'widow makers', with good cause. They are potential killers. Never design any electrical system that risks grid-level AC power running through exposed connectors. Your life, and the lives of the people around you are worth far more than saving a small amount of money.

COMPONENTS FOR STAND-ALONE SYSTEMS

If you are designing a stand-alone system, there is a lot more design work and planning involved than there is for a similarly-sized grid-tie system. It is more critical to make sure your stand-alone system works. A grid-tie system will not let you down if you do not generate enough energy, a stand-alone system will.

As well as considering and planning all the physical side of fitting the solar panels, routing the cabling and handling the safety aspects, you will also need to consider the voltage that your system will run at and design a battery system to store your energy.

CALCULATE YOUR OPTIMUM VOLTAGE

Solar panels and batteries are normally both 12 volts, so logically you would think that it would make the most sense to run your system at 12 volts.

For small systems, you would be right. However, there are some limitations of 12-volt systems. Therefore, we now need to identify the optimum voltage for your system.

If you are still not comfortable with volts, watts, currents and resistance, now would be a good time to re-read Chapter 2: *A Brief Introduction to Electricity*.

Voltages and currents

Current is calculated as watts divided by volts. When you run at low voltages, your current is much higher than when you run at higher voltages.

Take a normal household low-energy light bulb as an example. A 12W light bulb running from grid-level voltages is consuming 12 watts of power per hour. The current required to power this light bulb at 230 volts is 0.05 amps (12W ÷ 230V = 0.05 amps) and at 110 volts is 0.1 amps (12W ÷ 110v = 0.1 amps).

If you run the same wattage light bulb from a 12-volt battery, you are still only consuming 12 watts of power per hour, but this time the current you require is 1 amp (12W ÷ 12V = 1 amp).

If you run the same wattage light bulb from a 24-volt battery, you halve the amps. You now only require ½ amp (12W ÷ 24V = ½ amp).

"So what?" I hear you say. "Who cares? At the end of the day, we're using the same amount of energy, whatever the voltage."

The issue is resistance. Resistance is the opposition to an electrical current in the material the current is running through. Think of it as friction on the movement of electrons through a wire. If resistance is too high, the result is power loss. By increasing your voltage, you can reduce your current and thereby reduce energy loss.

You can counter the resistance by using thicker cabling, but you soon get to the point where the size of the cabling becomes impractical. At this point, it is time to change to a higher voltage.

What voltages can I run at?

It used to be the case that your whole system – both the energy generation and the energy usage – had to run at the same voltage. For example, if you had a string of four 12V solar panels, your solar array would run at 48V, and your battery and inverter had to run at 48V as well.

Whilst this is still often the case, it no longer has to be so. You can now purchase solar controllers that can take a string of solar panels that runs at a higher DC voltage and then steps this down to a lower voltage for charging one or more batteries at 12V, 24V or 48V. The maximum DC voltage from the solar array can be as high as 150V, depending on the solar controller used.

This allows you the opportunity to use the larger capacity, higher voltage solar panels usually found on grid-tie installations and use them in a stand-alone application.

For either a stand-alone or a grid fallback system, the most common voltages to run your batteries at are 12 volts, 24 volts or 48 volts.

As a rule, the most efficient way to run an electrical circuit is to keep your voltage high and your current low. That is why the grid runs at such high voltages It is the only way to keep losses to a minimum over long distances.

However, you also need to factor cost into the equation. 12-volt and 24-volt systems are far cheaper to implement than higher voltage systems, as the components are more readily available, and at a lower cost. 12-volt and 24-volt devices and appliances are also easily available, whereas 48-volt devices and appliances are rarer.

If you are designing and building your own system, it is unusual to go beyond 48 volts for the battery storage on stand-alone systems. Whilst you can go higher, inverters and controllers that work at other voltages tend to be extremely expensive and only suitable for specialist applications.

How to work out what voltage you should be running at

Your choice of voltage is determined by the amount of current (amps) that you are generating with your solar array or by the amount of current (amps) that you are using in your load at any one time.

To cope with bigger currents, you need bigger cabling and a more powerful solar controller. You will also have greater resistance in longer runs of cabling, reducing the efficiency of your system, which in turn means you need to generate more power.

In our system, we are proposing a 12m (40 feet) long cable run from the solar array to the house, plus cabling within the house.

Higher currents can also reduce the lifespan of your batteries. This should be a consideration where the current drain or charge from a battery is likely to exceed $1/10^{th}$ of its amp-hour rating.

We will look at battery sizing later, as current draw is a factor in choosing the right size of battery. It may be that you need to look at more than one voltage option at this stage, such as 12-volt and 24-volt, and decide which one is right for you later in the design process. Finally, if you are planning to use an inverter to convert your battery voltage to a grid-level AC voltage, 12-volt inverters tend to have a lower power rating than 24-volt or 48-volt inverters. This can limit what you can achieve purely with 12 volts.

To solve these problems, you can increase the voltage of your system. If you double the voltage, you halve your current.

There are no hard and fast rules on what voltage to work on for what current, but typically, if the thickness of cable required to carry your current has a cross section greater than 6mm² (or an AWG of 9 or below), it is time to consider increasing the voltage. I will explain how to calculate the required thickness of cable shortly.

HOW TO CALCULATE YOUR CURRENT

As explained in Chapter 2, it is very straightforward to work out your current. Current (amps) equals power (watts) divided by volts:

Power ÷ Volts = Current

P ÷ V = I

Go back to your power analysis (our Example 1 is on page 52) and add up the amount of power (watts) your system will consume if you switch on every electrical item at the same time. In the case of our holiday home, if I had everything switched on at the same time, I would be consuming 169 watts of electricity.

Using the holiday home as an example, let us calculate the current based on both 12 volts and 24 volts, to give us a good idea of what the different currents look like.

Using the above formula, 169 watts divided by 12 volts equals 14.08 amps. 169 watts divided by 24 volts equals 7.04 amps.

Likewise, we need to look at the solar array and work out how many amps the array is providing to the system. We need a 320-watt solar array. 320 watts divided by 12 volts equals 26.67 amps. 320 watts divided by 24 volts equals 13.33 amps.

CALCULATING CABLE THICKNESSES

I will go into more detail on cabling later, but for now, we need to ascertain the thickness of cable we will need for our system.

For our holiday home, we need a 12m (40 feet) cable to run from the solar controller to the house itself. Inside the house, there will be different circuits for lighting and appliances, but the longest cable run inside the house is a further 10m (33 feet).

That means the longest cable run is 22m (72½ feet) long. You can work out the required cable size using the following calculation:

$$(L \times I \times 0.8) \div V = CT$$

L		Cable length in metres (one metre is 3.3 feet)
I		Current in amps
V		System voltage (e.g. 12V or 24V)
CT		Cross-sectional area of the cable in mm^2

So calculating the cable thickness for a 12-volt system:

$$(22m \times 14.08A \times 0.8) \div 12V = 20.65mm^2$$

Here is the same calculation for a 24-volt system:

$$(22m \times 7.04A \times 0.8) \div 24V = 5.15mm^2$$

And just for sake of completeness, here is the same calculation for a 48-volt system:

$$(22m \times 3.52A \times 0.8) \div 48V = 1.63mm^2$$

Converting wire sizes:

To convert cross-sectional area to American Wire Gauge or to work out the cable diameter in inches or millimetres, use the following table:

Cross- Sectional Area (mm²)	American Wire Gauge (AWG)	Diameter (inches)	Diameter (mm)
107.16	0000	0.46	11.68
84.97	000	0.4096	10.4
67.4	00	0.3648	9.27
53.46	0	0.3249	8.25
42.39	1	0.2893	7.35
33.61	2	0.2576	6.54
26.65	3	0.2294	5.83
21.14	4	0.2043	5.19
16.76	5	0.1819	4.62
13.29	6	0.162	4.11
10.55	7	0.1443	3.67
8.36	8	0.1285	3.26
6.63	9	0.1144	2.91
5.26	10	0.1019	2.59
4.17	11	0.0907	2.3
3.31	12	0.0808	2.05
2.63	13	0.072	1.83
2.08	14	0.0641	1.63

1.65	15	0.0571	1.45
1.31	16	0.0508	1.29
1.04	17	0.0453	1.15
0.82	18	0.0403	1.02
0.65	19	0.0359	0.91
0.52	20	0.032	0.81
0.41	21	0.0285	0.72
0.33	22	0.0254	0.65
0.26	23	0.0226	0.57
0.2	24	0.0201	0.51
0.16	25	0.0179	0.45
0.13	26	0.0159	0.4

From these figures you can see the answer straightaway. Our cable lengths are so great that we cannot practically run our system at 12 volts. The nearest match for 20.65mm^2 cables is 21.14mm^2. This is AWG 4 cable, with a cable diameter of 5.19mm. Cable this size is thick, heavy, inflexible, hard to source and very expensive.

This means we would need to lay extremely thick AWG 4 cables from the solar array and around our house to overcome the resistance. This would be expensive, inflexible and difficult to install.

Realistically, due to cable sizing, we are going to need to use either 24 volts or 48 volts for our solar electric system.

MIXING AND MATCHING SOLAR PANELS

When specifying your solar array, you should keep to one type of panel rather than mixing and matching them. If you want a 100-watt array, for example, you could create this with one 100-watt solar panel, two 50-watt solar panels or five 20-watt solar panels.

If you do wish to use different solar panels in your array, you can do so by running two sets of panels in parallel with each other and either connecting them into a controller that can handle more than one feed, or by using more than one controller. This can be a useful way of creating the right wattage system, rather than spending more money buying bigger solar panels that generate more power than you need.

The result is slightly more complicated wiring, but it is often a more cost-effective solution to do this than to buy a larger capacity solar array than you need. There are two different ways of connecting two solar panels of different sizes to the same system. The next two diagrams show systems running at 12 volts, using two different sized panels to create a 140-watt peak system:

Single Controller system

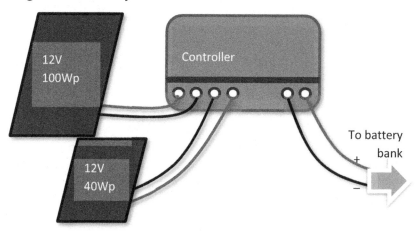

This system uses a single controller. The controller has two separate input feeds. The two solar panels work independently of each other and the controller handles the mismatch in power output. This is the best solution for handling mismatched solar panels.

Twin controller system

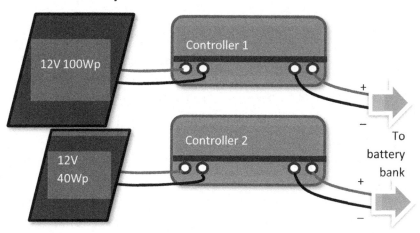

This system uses two controllers. The second controller is used to provide additional power to the batteries for the smaller solar panel. Both controllers charge the same battery bank.

These solutions are effective if you are planning to start small and add to your solar energy system when needs and budget allows. It means that you can collect an assortment of solar panels over time and put them to good use within your one solar energy system.

If you end up with two different makes of solar panels with identical ratings, put them on their own separate circuits. Solar panels from different manufacturers are not all identical in their operating voltages or performances, so putting two different makes and models of solar panel on the same circuit is likely to compromise the performance of both panels, even if the specification of the two panels is similar.

If you buy multiple controllers for your solar energy system, to handle different makes, models and sizes of solar panel, you only need one main controller to handle the *power output* from the batteries. Your other controllers can be much cheaper and simpler pieces of equipment as they are only handling the power feed into the batteries. If you wish, you can even use a simple solar regulator that simply cuts off charging when the batteries are full, although these are usually not as efficient as a proper solar inverter.

As discussed on page 129, not all solar panels are 12-volt panels. Many solar panels are now designed predominantly for grid-tie installation only and are available in many different voltage configurations. Solar panels with voltage ratings of up to 120 volts are on the market, although the most common are 12-volt and 24-volt panels.

In the past, the voltage of your solar array had to match the voltage of your battery system. If you had a 12-volt battery, you had to have a 12-volt solar panel to charge it. Likewise, if you had a 48-volt battery, you had to have several solar panels connected in series to match the 48-volt battery system.

That is still the case with many solar controllers. However, it is now possible to buy controllers that can handle much higher voltages from a solar array, and convert this to the lower voltage on a battery system. This gives you far greater flexibility when designing your system: you can use a few larger, higher voltage solar panels to produce the power you need, then use a solar controller to drop the voltage down to a suitable voltage for your battery pack.

Above: Victron Energy do a range of solar charge controllers that can work with high voltage solar arrays and reduce the voltage to a lower level for charging batteries. For example, the MPPT 150/45 controller shown above can work with a solar array of up to 150V/50A and can drop the voltage down to either 12, 24, 36 or 48-volts for the battery bank.

These new controllers work well for stepping down the voltage. For instance, you could have a solar array that produces a peak of 100–150 volts, charging up a 12-volt or 24-volt battery bank. However, these controllers cannot work the other way around: you cannot use them to charge a 24-volt battery pack from a single 12-volt solar panel, for example.

If your stand-alone solar electric system runs at a voltage other than 12 volts, you can either install multiple solar panels to boost the system voltage, or choose higher-voltage solar panels. For example, if you wanted a 200-watt 24-volt solar array, you could achieve this in various ways, including:

- Using two 12-volt, 100-watt solar panels, connected in series
- Using one 24-volt, 200-watt solar panel

When choosing a solar array, you need to consider:

- The physical size: will it fit into the space available?
- The support structure: ready-made supports may only fit certain combinations of panels
- How much cabling you will need to assemble the array
- The system voltage

BATTERIES

There are several different options when it comes to batteries, and plenty of specialist battery suppliers who can advise you on the best options for your solar installation. The two main battery technologies you will come across are lead-acid and lithium.

Lead acid batteries are the more established of the two technologies. Many people write them off as being out-of-date, whereas in fact, lead acid batteries have evolved significantly over the past ten years and remain the best option for most people.

Lithium batteries are much more expensive than lead acid batteries, but they often have a longer lifespan and are much more efficient in terms of energy storage. However, they are far more volatile than lead acid batteries. They require specialized charging and must never be allowed to overheat. Get things wrong with a lithium battery pack and you run the risk of of fire. If you do a search on Google or YouTube for 'lithium battery fire', you will very quickly see examples of what can go wrong with lithium-ion batteries when they overheat as a consequence of being charged, discharged or stored incorrectly.

I have worked with various different lithium-ion batteries over the past twenty years, starting with handheld computers in the mid-1990s and more recently with electric vehicles and solar installations. I have witnessed first-hand the difficulties that manufacturers have with building reliable lithium-ion battery systems.

Consequently, if you are considering going the lithium-ion route, you should buy a complete battery management solution, comprising of the solar charge controller, the batteries and the inverter. Do not mix and match components and do not attempt to build your own lithium ion battery pack from individual lithium-ion cells.

Buying a complete battery pack

Over the past two years, several companies have begun offering complete battery packs for solar energy systems. Products such as the Tesla Powerwall have received a lot of interest in solar energy storage and new products are being launched all the time.

Most of these battery systems are expensive, proprietary systems consisting of a battery pack, the control electronics and their own inverter. They are designed to work in conjunction with a grid-tie system. They tend to be sold on the advantages of being able to use your own electricity in the evenings as well as during the day, or on the peace of mind of offering a power backup in case of a power failure.

Whilst they are good in their own way, most of these systems are relatively inflexible. You have a battery pack of a fixed size and you are limited to the inverter provided. Most of the battery systems have a storage capacity in the region of 4kWh and a maximum power drain in the region of 600 – 800 watts.

These systems are designed to complement a grid-tie system. They charge up during the day and allow the energy to be used during the evening. In many cases, the battery systems are electrically separate from the solar array, charging up from an AC power source and releasing power through their own inverter. Most offer the facility to provide power in the event of a power cut, although often this requires a manual change-over. None of the mainstream solutions are suitable for off-grid systems.

Their inflexibility causes some real world issues that reduces their benefits for many people:

- Batteries usually take precident over other electrical demands within the home. If the batteries need recharging during the daytime, they will charge up from the available solar energy before other electrical demands are considered. Most battery suppliers claim that their batteries only charge up when electrical demand is low, but, if there is insufficient solar energy to both charge the batteries and run the electrics within the home over a period of a day, the batteries will take precident.
- Batteries can only charge up if there is sufficient energy being produced to charge up the batteries. This is not an issue with large solar arrays in sunny climates such as California, but in the United Kingdom, for example, where most residential solar arrays are rated as 4kWp or less, there is often insufficient energy being produced by the solar array during the winter to recharge a 4kWh battery pack on a daily basis. This means that the batteries may not receive a full charge every day during the winter, and the solar energy being produced by your system is exclusively being used to charge the batteries on these days.
- The built-in inverter can often only take a power demand of between 600 and 800 watts. Use more than this at any one time and you start taking electricity from the grid to make up the shortfall. So if you want to boil a

kettle, cook in an electric oven or on an electric hob or power an electric fan heater, you will not be using your battery system to provide all the power.

Commercially available battery systems are currently very expensive, often three to four times the cost of designing and building your own. Economically, their benefits are rarely justifiable unless there are significant additional grants available to help their adoption. Such benefits are available in California, with the Self-Generation Incentive Program (SGIP) and Hawaii, with their self supply tariff.

Designing and building your own battery pack

There are a lot of choices when choosing a battery system. Traditionally, lead acid batteries were used for energy storage, whilst more recently lithium-ion batteries have become more popular, thanks to their lighter weight, more compact size and more flexible charging and discharging characteristics.

Whatever battery technology you decide upon, battery systems must be carefully designed in order to get the best performance. Poorly designed battery systems may work fine when first installed, but are more likely to fail prematurely, leaving you with an unreliable system and the cost of replacing failing batteries.

Lead acid batteries usually come as 2-volt, 6-volt or 12-volt batteries, whilst lithium batteries usually come as 3.2V, 12V, 24V or 48V packs. Batteries can be connected in series to increase the voltage, or in parallel to keep the same voltage but increase the capacity.

The capacity of a battery is measured in amp-hours. The amp-hour rating shows how many hours the battery will take a specific drain: for instance, a 100-amp-hour battery has a theoretical capacity to power a 1-amp device for 100 hours, or a 100-amp device for 1 hour.

I say *theoretical*, because the reality is that all batteries provide more energy when discharged slowly. A 100-amp-hour battery will often provide 20–25% less power if discharged over a five-hour period, compared to discharge over a twenty-hour period.

No battery should be run completely flat. A minimum of 20% state of charge (SOC) should always be maintained in a battery to ensure the battery is not damaged. For best overall battery life, you should design your system so that the battery charge rarely goes below 50%.

Incidentally, manufacturers use one of two different measurements to show the amount of energy remaining in a battery. Some manufacturers give a 'state of charge' percentage, where 20% SOC means there is 20% energy remaining within the battery. Others give a 'depth of discharge' percentage, where 20%DOD means that 20% of the energy has been taken out of the battery (i.e. there is 80% energy left in the battery). When you are reading battery specifications, make sure you know which measurement is being referred to.

Types of lead acid batteries

There are four types of lead acid battery:

- 'Wet' batteries require checking and topping up with distilled water
- Lead Carbon batteries, that require no maintenance and provide superior performance, but at a higher cost
- AGM batteries require no maintenance but have a shorter overall life and are less suitable for demanding applications
- Gel batteries are also maintenance-free and provide a good overall life. They can be placed on their side or used on the move

In the past, most installers have recommended industrial quality 'wet' batteries for all solar installations. These provide the best long-term performance and the lowest cost. Often called *traction* batteries (as they are heavy-duty batteries used in electric vehicles), the latest designs can often have a lifespan of 15–20 years for a solar installation.

However, one of the latest innovations with lead acid batteries is the introduction of carbon to the negative plates inside the battery. These lead carbon batteries (also known as Advanced Lead Carbon or ALC batteries) have significantly extended the lifespan and performance of lead acid batteries, making them comparable to lithium batteries. Like lithium batteries, they can be charged and discharged more rapidly than other lead acid batteries. Unlike lithium batteries, they continue to perform well at sub-zero temperatures and do not require expensive and complicated thermal management. The disadvantage of lead carbon batteries is that they have a higher voltage drop-off as the charge drops, which can cause problems for some applications.

A lower cost option to the industrial-quality traction battery is the leisure battery, as used in caravans and boats. These are typically AGM batteries. Their lifespan is considerably shorter than traction batteries, often requiring replacement after 5–6 years and significantly less in intensive applications.

The final option is the gel battery. These have the benefit of being entirely maintenance-free. They are also completely sealed and do not emit hydrogen gas. In the past, gel batteries have not been particularly reliable in solar installations, tending to require replacement after 1–2 years. However, this is no longer the case, with the latest generation of long-lasting gel batteries providing a lifespan of 8–10 years for a solar installation. The price of these batteries has also dropped significantly.

Gel batteries can provide an excellent, zero-maintenance alternative to wet batteries for smaller applications and the latest gel batteries are suitable for higher-drain applications as well as smaller systems.

Not all battery makes are the same. From my experience, the very best battery manufacturers for solar energy installations are Trojan, Crown, Northstar and Enersys.

Types of Lithium-ion battery

The benefits of lithium-ion batteries over lead acid batteries are they are lighter, take up less space, can be regularly discharged to 20% state of charge and will not deteriorate as quickly as most lead acid batteries over time.

However, many of these advantages are diminishing when compared to the latest lead carbon batteries that can now match the lifetime of lithium-ion batteries at a significantly lower cost.

Like lead acid batteries, lithium-ion batteries also come in different forms, the most common ones are:

- Lithium-ion Cobalt Oxide ($LiCoO_2$), usually called Lithium Cobalt, are the cheapest batteries, often found in laptops and mobile phones. They have the shortest life and are not suitable for most solar energy storage applications.
- Lithium-ion Manganese Oxide ($LiMn_2O_4$), usually called Lithium Manganese can be charged and discharged much quicker than Lithium Cobalt and has a longer life.
- Lithium-ion Maganese Cobalt Oxide ($LiMn_2CoO_4$), usually called NMC batteries, and their close cousins the Lithium-ion Nickel Cobalt Aluminium Oxide ($LiNiCoAlO_2$), or NCA batteries, are the latest development of the Lithium Maganese batteries. They have a high power density, meaning the battery stores more energy than other batteries of the same size and weight. They are chemically more stable than Lithium Maganese batteries,

particularly when discharging rapidly. These are the most common batteries used for energy storage systems, and are also used by Tesla in their electric vehicles.

- Lithium-ion Iron Phosphate (LiFePO$_4$), usually called Lithium Phosphate, are heavier and bulkier than most other lithium-ion battery chemistries, but are also heavier duty batteries capable of providing larger amounts of power over short periods of time. They are less tolerant of low temperatures, however, performing poorly at close to freezing temperatures.

There are other lithium-ion battery technologies, including Lithium Titanate, Lithium Salt, Lithium Graphite and Lithium Air, but these technologies are either very specialist or still at the experimental stages.

You may also hear about Lithium Polymer batteries, often referred to as LiPo batteries. Lithium Polymer is not a chemistry by itself, rather it is a way of manufacturing the batteries in a pouch design. Most batteries advertised as Lithium Polymer batteries are Lithium Cobalt batteries.

The biggest issue with all lithium-ion batteries is that they require careful battery management. Lithium-ion batteries must be paired with a suitable battery management system (BMS) that manages the current flows in and out of the battery, monitors internal battery temperatures and ensures the safety of the overall system.

Lithium-ion batteries can easily be destroyed or damaged by incorrect charging or discharging. Most of them become unstable if they overheat, which can happen by charging or discharging the batteries too quickly, or by overcharging them when they are close to being fully charged, or by charging them at too high a voltage, or by through a short-circuit. An unstable lithium-ion battery can catch fire or explode. Because all lithium-ion batteries contain oxygen, putting out a lithium-ion fire is difficult, and the fire can very quickly spread from one battery to the whole pack.

Do not be tempted to mix and match lithium-ion batteries and charge controllers. The charge controller you purchase must have the capability of charging lithium batteries and working with the BMS in the battery. Furthermore, some of the cheap lithium-ion battery systems that are often found on eBay often do not come with a BMS of their own or have an inadequate BMS that does not protect the batteries sufficiently, particularly if they are being charged using solar rather than from the manufacturers own battery charger.

If you are considering lithium-ion batteries, it is vitally important that you chose a battery system with a good quality BMS, plus a solar controller that can operate with the lithium-ion battery pack you choose. It is common for solar controller manufacturers to either provide their own lithium-ion battery packs or provide a list of approved lithium ion batteries to work with.

If considering a lithium-ion battery solution, you also need to ensure you have the right environment to keep them in: they must be kept in dry conditions with an ambient temperature of between 5° and 35°C (41°–95°F).

Battery recycling

Batteries do not last forever, and they must not be disposed of with other waste. Lithium-ion batteries, in particular, require special processing and may explode or catch fire if crushed for landfill in a recycling facility.

Developments in recycling lithium-ion batteries has been slow over the past ten years. Presently only 20-30% of the contents of a lithium-ion battery can easily be recycled and the cost of recycling is currently uneconomic. You may have to pay to have your old lithium-ion batteries recycled.

Lead acid batteries, however, are widely recycled and around 95-98% of a lead acid battery can be recycled to create new lead acid batteries. The value of the materials inside a lead acid battery are also significant and an individual scap battery can often be sold for £15-20 ($20-28).

Battery configurations

You can use one or more batteries for power storage. Like solar panels, you can wire your batteries in parallel to increase their capacity or in series to increase their voltage.

It is important that you to use the same specification and size of batteries to make up your battery bank. Mixing battery capacities and types will mean that some batteries will never get fully charged and some batteries will get discharged more than they should be. As a result, mixing battery capacities and types can significantly shorten the lifespan of the entire battery bank.

Battery lifespan

Batteries do not last forever, and at some stage in the life of your solar electric system, you will need to replace them. Obviously, we want to have a battery system that will last as long as possible and so we need to find out about the lifespan of the batteries we use.

There are two ways of measuring the lifespan of a battery, both of which tell you something different about the battery.

- Cycle Life is expressed as a number of cycles to a particular depth of discharge
- Life in Float Service shows how many years the battery will last if it is stored, charged up regularly, but never used

Cycle life

Every time you discharge and recharge a battery, you *cycle* that battery. After a number of cycles, the chemistry in the battery will start to break down and eventually the battery will need replacing.

The cycle life will show how many cycles the batteries will last before they need to be replaced. The life is shown to a 'depth of discharge' (DOD), and the manufacturers will normally provide a graph or a table showing cycle life verses the depth of discharge.

Typical figures that you will see for cycle life may look like this:

	Lead Carbon battery	Lead Acid Gel battery	Lithium-ion battery
20% DOD	9,000 cycles	2,000 cycles	9,000 cycles
40% DOD	5,500 cycles	1,400 cycles	6,000 cycles
50% DOD	4,500 cycles	1,200 cycles	5,000 cycles
80% DOD	1,800 cycles	700 cycles	2,000 cycles

As you can see, the battery will last much longer if you keep your depth of discharge low. For this reason, it can often be better to specify a larger battery, or bank of batteries, rather than a smaller set of batteries. Most experts recommend that you install enough batteries to ensure that your system does not usually discharge your batteries beyond 50% of their capacity if using lead acid batteries, or 75% of their capacity if using lithium-ion.

There are other benefits of large battery banks, too:

- You have more flexibility with your energy usage: If you need to use more electricity for a few days than you originally planned for, you know you can do this without running out of energy.
- Batteries degrade over time: a battery nearing the end of its life may only have 70-80% of the capacity of a new battery. By over-specifying the battery pack, it will not let you down in years to come as the batteries age.

Holdover

When considering batteries, you need to consider how long you want your system to work while the solar array is not providing any charge at all. This time span is called *holdover*.

Unless you live inside the Arctic or Antarctic Circles (both of which provide excellent solar energy during their respective summers, incidentally), there is no such thing as a day without sun. Even in the depths of winter, you will receive some charge from your solar array every day.

You may find there are times when the solar array does not provide all the energy you require. It is therefore important to consider how many days holdover you want the batteries to be able to provide power for, should the solar array not be generating all the energy you need.

For most applications, a figure of between three days and five days is usually sufficient.

In our holiday home, we are deliberately not providing enough solar energy for the system to run 24/7 during the winter months. During the winter, we want the batteries to provide enough power to last a long weekend. The batteries will then be recharged when the holiday home is no longer occupied and the solar panel can gradually recharge the system.

For this purpose, I have erred on the side of caution and suggested a five-day holdover period for our system.

Calculating how long a set of batteries will last

Calculating how long a set of batteries will last for your application is not a precise science. It is impossible to predict the number of discharges, as this will depend on the conditions the batteries are kept in and how you use the system over a period of years.

Nevertheless, you can come up with a reasonably good prediction for how long the batteries should last. This calculation will allow you to identify the type and size of batteries you should be using.

First, write down your daily energy requirements. In the case of our holiday home, we are looking at a daily energy requirement of 695 watt-hours.

Then, consider the holdover. In this case, we want to provide five days of power. If we multiply 695 watt-hours a day by 5 days, we get a storage requirement of 3,475 watt-hours of energy.

Batteries are rated in amp-hours rather than watt-hours. To convert watt-hours to amp-hours, we divide the watt-hour figure by the battery voltage.

If we are planning to run our system at 12 volts, we divide 3,475 by 12 to give us 290 amp-hours at 12 volts. If we are planning to run our system at 24 volts by wiring two batteries in series, we divide 3,475 by 24 to give us 145 amp-hours at 24 volts.

We do not want to completely discharge our batteries, as this will damage them. So we need to look at our cycle life to see how many cycles we want. We then use this to work out the capacity of the batteries we need.

During the spring, summer and autumn, we are expecting the solar array to recharge the batteries fully every single day. It is unlikely that the batteries will be discharged by more than 10–20%. However, during the winter months, we could have a situation where the batteries get run down over a period of several days before the solar panels get a chance to top the batteries back up again.

So, for four months of the year, we need to take the worst-case scenario where the batteries may get discharged down to 80% depth of discharge over a five-day period and then recharged by the solar array.

If I use lithium-ion or lead carbon batteries, the batteries will allow us to do this around 2,000 times before they come to the end of their useful life. If I were to use lead acid gel batteries, the cycle life is closer to 700 times.

2,000 cycles multiplied by 5 days = 10,000 days = 333 months, whilst 700 cycles multiplied by 5 days = 3,500 days = 116 months. As this scenario will only happen during the four months from November to February, lead carbon or lithium-ion batteries will last us for around forty years before reaching the end of their cycle life, whilst the gel batteries will last almost twenty years.

In reality, the *Life in Float Service* figure (i.e. the maximum shelf-life) for batteries is likely to be around fifteen to twenty years, which means that in this instance they will fail before they reach their cycle life.

Based on our energy requirements of 145 amp-hours at 24 volts, and a maximum discharge of 80%, we can calculate that we need a battery capacity of 145 ÷ 0.8 = 181.25 amp-hours at 24 volts.

Second-hand lead acid batteries

There is a good supply of second-hand lead acid batteries available. These are often available as ex-UPS batteries (UPS = Uninterruptable Power Supplies) or ex-electric vehicle batteries.

Whilst these will not have the lifespan of new batteries, they can be extremely cheap to buy, often selling at their scrap value. If you are working to a tight budget and your power demands are not great, this is a very good way to save money, particularly as you will be able to sell them yourself for scrap value when it is time to replace them.

Bear in mind that the technology in lead acid batteries have improved massively over the past three years, which means that older batteries will have a much shorter cycle life and overall lifespan when compared with the latest technology batteries. However, if you can get second-hand batteries for close to scrap value, you are effectively getting energy storage almost for free, and that can be hard to argue against!

Do not 'mix and match' different makes and models of batteries. Use the same make and model of battery throughout your battery bank. I would also advise against using a mixture of new and used batteries. This is a false economy as the life of your new batteries may be compromised by the older ones. If you are considering second-hand batteries, try and find out how many cycles they have had and how deeply they have been discharged. Many UPS batteries have hardly been cycled and have rarely been discharged during their lives.

If buying ex-electric vehicle batteries, remember these have had a very hard life with heavy loads. However, ex-electric vehicle batteries can continue to provide good service for lower- demand applications. If your total load is less than 1kW and you have enough batteries to ensure they discharge slowly over several hours, these batteries can provide good service.

If possible, try and test second-hand batteries before you buy them. Ensure they are fully charged up, and then use a battery load tester on them to see how they perform.

If your second-hand batteries have not been deep cycled many times, the chances are they will not have a very long charge life when you first get them. To 'wake them up', connect a solar controller or an inverter to them and put a low-power device onto the battery to drain it to around 20% state of charge. Then charge the battery up again using a trickle charge and repeat.

After three deep cycles, you will have recovered much of the capacity of your second-hand batteries.

If using second-hand batteries, expect them to provide half of their advertised capacity. So if they are advertised as 100-amp-hour batteries, assume they will only give you 50 amp-hours of use. In the case of ex-electric vehicle batteries, assume only one-third capacity.

The chances are, they will give you much more than this, but better to be happy with the performance of your second-hand batteries than to be disappointed because they are not as good as new ones.

Second hand lithium-ion batteries

Second hand lithium-ion batteries are also available, but these should be avoided unless you really understand what you are doing, or working with a lithium-ion battery professional who can configure the whole battery system for you. As with lead acid batteries, buying second hand lithium ion batteries can save money, but it is an extremely specialist area and not to be tackled without in-depth battery system knowledge.

Building your battery bank

Because we are running our system at 24 volts, we will need two 12-volt batteries connected in series to create our battery bank.

We therefore need two 12-volt batteries of 181.25 amp-hours each to create the desired battery bank.

It is unlikely that you are going to find a battery of exactly 181.25 amp-hours, so we need to find a battery that is *at least* 181.25 amp-hours in size.

When looking for batteries, you need to consider the weight of the batteries. A single 12-volt lead acid battery of that size will weigh in the region of 60kg (over

130 pounds), whilst a lithium-ion battery will weigh around 35-40kg (77-88 pounds).

Safely moving a battery of that size is not easy. You do not want to injure yourself in the process. A better solution may be to buy multiple smaller batteries and connect them together to provide the required capacity.

Because lead acid batteries contain acid, they should be installed in a battery tray, so that any acid leaks may be contained. In Australia and Canada, regulations state that all batteries must be enclosed in a ventilated, lockable and vermin-proof enclosure.

As it is not possible to buy 181.25 amp-hour batteries, I have decided to use four 100-amp-hour 12-volt batteries, giving me a battery bank with a total capacity of 200 amp-hours at 24 volts.

12-volt, 100-amp-hour batteries are still not lightweight. They can easily weigh 30kg (66 pounds) each, so do not be afraid to use more, lighter-weight batteries, if you are at all concerned.

To build this battery bank, you can use four 100-amp-hour, 12-volt batteries, with two sets of batteries connected in series, and then connect both series in parallel, as shown over the page:

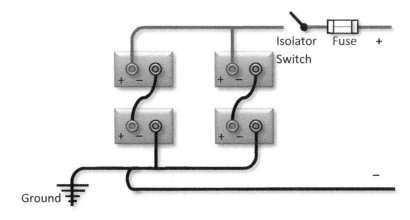

Four 100-amp-hour 12V batteries. I have paired up the batteries to make two sets of 100-amp-hour 24V batteries, and then connected each pair in parallel to provide a 200-amp-hour capacity at 24 volts.

If I were putting together a 12-volt battery system instead of a 24-volt battery system, I could wire together multiple 12-volt batteries in parallel to provide the higher capacity without increasing the voltage:

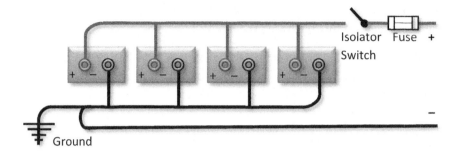

Four 100-amp-hour 12V batteries connected in parallel to provide a 400-amp-hour 12V battery bank.

Battery safety

Battery safety is an area that must be carefully considered for your installation. This is particularly true of lead acid batteries, as they contain highly corrosive acid and they vent explosive hydrogen gas during their charging cycles. 'Wet' lead acid traction batteries regularly need their water levels topped up as well, which can lead to accidental acid splashes.

In addition, there is the very real risk of shorting a battery by accidentally dropping a spanner or wrench and shorting it across two terminals. Do this and the spanner can turn red hot in less than a second, as well as potentially causing permanent damage to your battery.

So as well as the handling issues of moving heavy, bulky batteries into position, you need to carefully consider the battery maintenance aspects and consider how hydrogen can safely vent and be expelled from the area.

I will go into more detail about handling batteries and ensuring you don't accidentally create short circuits when building your battery pack during the chapter on installation. However, the aspect of hydrogen venting has to be covered in the design phase of your system, so I shall cover this now.

Venting Hydrogen from Lead Acid batteries

Different batteries vent different amounts of hydrogen. Sealed lead acid batteries, such as all gel and most AGM batteries, should never vent hydrogen unless they become damaged, at which point a small amount of hydrogen may vent. Wet lead acid batteries will vent hydrogen every time they charge up, particularly as the batteries near being fully charged. Whilst the amount of hydrogen being vented will vary depending on the size of the battery, the exact chemistry and the amount of current you are using, a rough rule of thumb is to assume a worst case of 0.011 cubic metres of hydrogen could be released from every kWh of 'wet' battery storage per hour during the final three hours of the charging cycle. One kWh of battery storage equals one 85Ah 12V battery or one 170Ah 6V battery.

Hydrogen cannot be seen, nor can it be smelt. But it is an explosive gas that can explode when it reaches 4% concentration in air. It is generally accepted that 1% concentration is the highest 'safe' level of hydrogen in an environment.

Thankfully, hydrogen is very easy to extract. As one of the smallest and lightest molecules, it will naturally rise up and away from the batteries. It will then dissipate quickly into the atmosphere. However, there can be a problem with hydrogen getting trapped in buildings where it can mass in the eves of a roof or get trapped in the corner of a room.

If you only have one or two batteries in a large room, you need not worry as the amount of hydrogen that can be released from the batteries is too small to worry about. However, if you have a larger battery bank, you need to consider how to vent hydrogen outside the building.

There are different ways of doing this: natural convection, creating a passive ventilation system with no moving parts, or a vented fan solution. Vented fan solutions cause their own problems, in particular what happens if the fan fails, and the turbulence caused by the fans can create unexpected hydrogen pockets elsewhere in the battery enclosure. For these reasons, a passive ventilation system is usually recommended for lead acid battery storage solutions.

The principles of natural convection

Natural convection is the process of moving liquid or gas simply through taking advantages of the different densities of the liquid or gas, either due to different gases being present or through temperature differences within the gas.

For example, we all know that warm air rises and cold air drops to the bottom of a container. Because we are working with different gases, we can use the same principles with extracting hydrogen from batteries.

Hydrogen is significantly lighter than air and will always rise to the top of a chamber. Heavier gases, such as air, will sink downwards below the hydrogen.

Where the temperature of the hydrogen and the temperature of the surrounding air is the same, hydrogen will move upwards at the rate of 20 metres per second (65 feet per second).

Using natural convection to extract hydrogen

The best approach for using natural convection is to build a battery box with a sloped, hinged lid. The underside of the lid should be flat with no traps. At the very top of the battery box, a PVC pipe can be fitted into the lid creating a chimney. This chimney can then be vented to the outside.

As hydrogen is vented from the batteries, it will immediately rise to the top of the battery box and directed by the slope of the lid towards the PVC pipe.

The diameter of the PVC pipe can be calculated based on the speed of the hydrogen and the estimated worst case for hydrogen production from the batteries. The calculation for the inside pipe diameter is:

$$\text{Extraction pipe diameter in centimetres} = \sqrt{\frac{0.011 \times Total\ kWh\ of\ batteries \times 1000}{\pi \times 20m/s}}$$

To convert millimetres to inches, divide the answer by 2.54.

If you only have 1 kWh of battery storage, the answer for your calculation is that you need a pipe with a minimum internal diameter of 0.42cm ($^1/_6$ inch). If you have 8kWh of battery storage, the calculation suggests that you need a pipe with a minimum internal diameter of 1.19cm (just under ½ inch).

There is a safety issue with such small extraction pipes: they can easily get blocked, and the gas release can be hampered by the friction against the pipe surface. Consequently, it is usually recommended that you have a minimum extraction pipe size of at least 1.25cm (½ inch) for any battery box extraction pipe and that you increase the size of the pipe by at least 20% beyond the calculated value. If you have a grille on the outside of your pipe, you will also need to increase the size of the pipe by at least 50% to account for the reduction in airflow created by the grille.

If you are extracting gas from inside the battery box, you need to replace it with something else: typically fresh air from the area surrounding the box. You will typically do this with vents near the bottom of the battery box. Because the flow of air into the box will be slower than the release of hydrogen out of it, the size of the intake vents needs to be correspondingly larger:

$$\text{Intake vent diameter in centimetres} = \sqrt{\frac{0.011 \times \text{Total kWh of batteries} \times 1000}{\pi \times 1 m/s}}$$

Before you panic that you must work out all these calculations yourself, here is a table that shows the minimum intake and extraction pipe diameters for the battery sizes you are most likely to be working with:

Total Battery pack size (kWh)	Extraction pipe minimum internal diameter	Inlet vent minimum internal diameter
1kWh	0.42cm ($^1/_6$ inch)	1.87cm ($^3/_4$ inch)
2kWh	0.59cm (¼ inch)	2.64cm ($1^1/_{20}$ inch)
3kWh	0.73cm ($^5/_{16}$ inch)	3.24cm (1¼ inch)
4kWh	0.84cm ($^1/_3$ inch)	3.74cm (1½ inch)
6kWh	1.02cm ($^2/_5$ inch)	4.58cm ($1^4/_5$ inch)
8kWh	1.19cm (½ inch)	5.29cm ($2^1/_{10}$ inches)
10kWh	1.32cm ($^9/_{16}$ inch)	5.92cm ($2^1/_3$ inches)
20kWh	1.87cm ($^3/_4$ inch)	8.37cm ($3^1/_3$ inches)

Hydrogen Sensors and Alarms

Whilst venting hydrogen is not a big issue, the consequences of getting it wrong can be huge. If you have built a battery box with adequate ventilation, the likelihood of a hydrogen explosion is remote. Furthermore, most hydrogen explosions are tiny (rarely more than a quiet pop). However, a large quantity of hydrogen building up in a room or inside a battery enclosure can have catastrophic consequences.

To reduce this risk, if you are installing lead acid batteries inside a building, or in a boat or recreational vehicle, consider installing two hydrogen alarms – one at the top of the battery enclosure and one at the ceiling in the room where the batteries are installed. These hydrogen alarms will trigger if the hydrogen concentration gets beyond 1% (hydrogen is not dangerous until it reaches 4% concentration in air). If the alarms trigger, this gives you plenty of time to disconnect the battery pack and ventilate the area before there is any real danger. As with a smoke alarm in your home, the likelihood you will ever need them is remote. But the peace of mind they can provide is huge.

SOLAR CONTROLLER

The solar controller looks after the batteries and stops them either being overcharged by the solar array or over-discharged by the devices running off the batteries.

Many solar controllers also include an LCD status screen so you can check the current battery charge and see how much power the solar array is generating. More expensive solar controllers may have the option of a WiFi or Bluetooth link and an app to run on your mobile phone, so you can monitor the performance of your system.

Your choice of solar controller will depend on five things:

- System voltage
- The type of batteries you are using (particularly important if you are planning to use lithium batteries)
- The current of the solar array (measured in amps)
- The maximum current of the load (measured in amps)
- The level of detail you require from the status display

All but the very cheapest solar controllers provide basic information on an LCD screen that allows you to see how much power you have generated compared to how much energy you are using and can also show the current charge stored in the battery. Some solar controllers include more detailed information that allows you to check daily how your power generation and usage compares.

Balancing the batteries

Another important function of a solar controller is to manage the charge in each battery and to ensure each battery is properly charged up.

As batteries get older, the charge of each battery will start to vary. This means that some batteries will charge and discharge at different rates to others. If left over time, the overall life of the batteries will deteriorate.

Intelligent solar controllers can manage these variations by balancing, or *equalizing*, the batteries they are charging. On most controllers, you need to manually activate a balance as part of a routine inspection.

Allow for expansion

When looking at solar controllers, it is worth buying one with a higher current rating than you need.

This allows you extra flexibility to add additional loads or additional panels to your solar array in the future without having the additional expense of replacing your solar controller.

Maximum power point tracking

More expensive solar controllers incorporate a technology called *maximum power point tracking* (MPPT). An MPPT controller adjusts the voltage being received from the solar array to provide the optimum voltage for charging the batteries without significant loss of watts from the voltage conversion.

If you have an MPPT controller, you can capture around 20% more of the power generated by the solar array compared to a more basic controller.

If you have less than 120W of solar panels, it can work out cheaper to buy extra solar panels rather than spend the extra money on an MPPT controller. However, prices continue to fall and, if you have the choice, a controller with maximum power point tracking is a worthwhile investment.

Ground fault protection

Many solar controllers include ground fault protection. In the case of a short from the solar array, a *Residual Current Device* (RCD) will cut off the current flow between the solar array and the controller, thereby averting the risk of damage to either the controller or the solar array.

In the United States and Canada, RCDs are also known as *Ground Fault Interrupters* (GFIs).

For anything larger than 100-watt solar panel systems, and for all systems mounted to a building, you need to incorporate a separate RCD/GFI into your system if you do not have ground fault protection built into your controller.

Backup power

Some controllers have the capability to start up an emergency generator if the batteries run too low and the solar array is not providing enough power to cope with the load. This can be a useful facility for sites where the system must not fail at any time, or for coping with unexpected additional loads.

Whilst this may not seem so environmentally friendly, many generators are now available that run on bio-diesel or bio-ethanol. Alternatively, you can use an environmentally friendly fuel cell system instead of a generator. These tend to run on bio-methanol or zinc and only emit water and oxygen.

Using multiple controllers

Sometimes it is desirable to have multiple controllers on your solar energy system. For instance, you may want to install solar panels in different locations or facing in different directions, or you may have mismatched solar panels that you want to use (see page 133 for a block diagram).

Multiple controllers are not an option with lithium-ion batteries but can be a useful solution with lead acid batteries. If you need to have multiple controllers, only one needs to have the expensive features such as battery balancing. The other controllers can be much simpler regulators that simply provide an additional charge to the batteries and switch off when the batteries are fully charged.

A better solution to having multiple controllers is often to have one controller that can handle feeds from more than one solar array. This can work well if you only need two strings of solar panels and can work with lithium-ion batteries.

INVERTERS

We are not using an inverter with our holiday home, but many solar applications do require an inverter to switch up the voltage to grid-level AC current.

An inverter for stand-alone systems is a different piece of equipment to a grid-tie solar inverter. With a grid-tie inverter, your power is feeding into the grid and must work in conjunction with the grid. The inverter connects directly to your solar panels and switches off when the solar panels no longer produce enough energy.

With a stand-alone system, your power is entirely separate from the grid. The inverter connects to your battery bank and switches off when the battery bank is running low on charge.

There are three things to consider when purchasing an inverter:

- Battery bank voltage
- Power rating
- Waveform

Battery bank voltage

Different inverters require a different input voltage. Smaller inverters, providing up to 3kW of power, are available for 12-volt systems. Larger inverters tend to require higher voltages.

Power rating

The power rating is the maximum continuous power that the inverter can supply to all the loads on the system. You can calculate this by adding up the wattages of all the devices that are switched on at any one time. It is worth adding a margin for error to this figure. Inverters will not run beyond their maximum continuous power rating for very long.

Most inverters have a peak power rating as well as a continuous power rating. This peak power rating allows for additional loads for very short periods of time, which is useful for some electrical equipment that uses an additional burst of power when first switched on (refrigeration equipment, for example).

As a general rule of thumb, go for a bigger power rating than you actually need. Inverters can get very hot when they get close to their maximum load for long periods of time. Many professionals recommend that you buy an inverter that has a continuous power rating that is at least one third higher than you plan to use.

Waveform

Waveform relates to the quality of the alternating current (AC) signal that an inverter provides.

Lower-cost inverters often provide a *modified sine wave* signal (sometimes advertised as a *quasi-sine wave*). More expensive inverters provide a *pure sine wave* signal.

Modified sine wave inverters tend to be considerably cheaper and also tend to have a higher peak power rating.

However, some equipment may not operate correctly with a modified sine wave inverter. Some power supplies, such as those used for laptop computers and portable televisions, may not work at all, while some music systems emit a buzz when run from a modified sine wave inverter.

These faults are eliminated with a pure sine wave inverter, which produces AC electricity with an identical waveform to the standard domestic electricity supply provided by the grid.

Installation options and operating environment

Small inverters with a continuous power rating of less than 3kW are lightweight units and are often simply placed on a shelf or a desk. Medium-sized inverters tend to be heavy units that need to be mounted securely on a wall. Larger inverters, rated at 10kW or above, may need to be bolted to a floor.

Many inverters generate a significant amount of heat, especially when running close to their rated output, and require good airflow around the unit.

Most off-grid inverters are designed to be installed inside. Outdoor inverters are available, but they are expensive and may be difficult to source. If you are looking at installing an inverter outdoors, check that the inverter is sealed against dust and water ingress, rated to at least IP64.

Overheating inverters is the number one reason for any solar system failing. As an inverter gets hotter, they provide less power, and if the temperature continues to rise they will eventually shut down to avoid permanent damage. When choosing an inverter, check its operating temperatures and consider how you can ensure your system remains within these limits.

Inverters should always be installed in a well-ventilated area, away from the ceiling and with a clearance around each side, the top and the bottom. They cannot be installed in a sealed cupboard. Some inverters have the option of an external heat sink, or temperature-controlled cooling fans to help keep the inverter cool.

If your inverter is producing more than around 500 watts of power, it is likely to make a very small amount of noise. This is typically a continuous low-level hum. This is usually only noticeable if the surrounding area is quiet. However, for this reason, inverters are not usually installed inside the living space in a home or in an office environment. Instead, consider installing your inverter in a garage, or on an outside wall of your building.

Occasionally, with larger inverters, the sound made by the inverter has been known to resonate with the wall, amplifying the sound and making it quite unpleasant to live with, even if the inverter is mounted to an outside wall or the wall of a garage. This is a very rare occurrence but is most likely to occur if you are planning to mount your inverter onto a wall made out of solid concrete. The solution is to dampen the mounting between the inverter and the wall or floor that the inverter is mounted onto. There is a very effective product called *Green Glue*, produced by the Green Glue Company (*www.GreenGlueCompany.com*) that is applied between the wall and the inverter. When it is compressed by remounting the inverter, the glue spreads out to form a sound isolation barrier that is particularly effective at blocking out low-resonance vibrations.

Bigger inverter systems

The sky is the limit with inverter sizes: I have personally worked with multiple inverter systems that can generate almost a megawatt in power. Three phase inverters are available, as are inverters that can work at much higher voltages – up to 33,000 volts in some cases. Some of these inverters are designed to work only in conjunction with the grid, whilst others can be 'grid forming', creating their own microgrid.

Many of these inverter systems are designed to work in parallel, so that you can scale them up as your requirements grow. Some systems, such as the Victron Quattro or SMA Sunnyboy inverter systems give you a huge amount of flexibility: you can configure your system so that it runs in conjunction with the grid, or forms its own microgrid if there is no grid connection. They can even dynamically switch between the two if you wish. You can use three inverters in parallel to create a three-phase output. You can use multiples of these inverters to increase the power generation from 3kW right up to 200kW and beyond. You can choose to export energy back to the grid or reserve it for your own use.

Whilst these inverter systems are not typically designed for home use, they do have uses in industry, or for specialist applications for backup power or supporting a weak or unreliable grid.

Ground fault protection

Most inverters now include ground fault protection. All inverters must always be grounded. If your chosen grid-tie inverter does not incorporate ground fault protection, you need to incorporate this into your system using a *Residual Current Device* (RCD). RCDs are known as *Ground Fault Interrupters* (GFIs) in the United States and Canada.

COMBINED SOLAR CONTROLLER AND INVERTERS

In the past couple of years, it has become possible to buy combined solar controller and inverters. Some of these are designed specifically for off-grid applications, others allow for both off-grid and grid-tied operation.

The benefit of these systems is that there is reduced wiring, and usually a single unit to set up and configure. There can often be a cost saving as well.

The disadvantage of a combined system is that you may be restricted in your choice of batteries or the upper voltage of your solar array. However, in practice, most of these combined systems have a good enough power input and battery configuration range to support most applications.

Victron, Growatt and Iconica all supply combined controller/inverters. I have personally used the Victron and Iconica solutions and found them very effective for off-grid applications.

CABLES

It is easy to overlook them, but cables have a vital part to play in ensuring a successful solar electric system. There are three different sets of cables that you need to consider:

- Solar array cables
- Battery cables
- Appliance cabling

Solar array cabling has already been discussed (see page 112). For stand-alone systems, the battery and appliance cabling also need to be correctly specified.

For all cabling, make sure that you always use cable that can cope with the maximum amount of current (amps) that you are planning to work with.

Remember that you may wish to expand your system at some point in the future. Use a higher ampere cable than you need to make future expansion as simple as possible.

Battery cables

Battery cables are used to connect batteries to the solar controller and to the inverter. They are also used to connect multiple batteries together.

Battery interconnect cable is available ready-made up from battery suppliers, or you can make them up yourself. You should always ensure that you use the correct battery connectors to connect a cable to a battery.

Appliance cabling

If you are using an inverter to run your appliances at grid-level voltage, you can use standard domestic wiring, wired in the same way as you would wire them for connection to domestic AC power.

If you are running cabling for 12-volt or 24-volt operation, you can wire your devices up using the same wiring structure as you would use for grid-level voltage, although you may need to use larger cables throughout to cope with the higher current.

In a house, you would typically have a number of circuits for different electrical equipment. One for downstairs lighting, one for upstairs lighting, one for an electric cooker and one, two or three for appliances, depending on the size of the house. This has the benefit of keeping each individual cable run as short as possible, as well as reducing the amount of current that each circuit needs to handle.

As we have already learnt, low-voltage systems lose a significant amount of power through cabling. The reason for this is that the current (amps) is much higher and the power lost through the cable is proportional to the square of the current. You therefore need to keep your cable runs as short as possible, especially the cable runs with the highest current throughput.

I have already mentioned how you can calculate suitable cable thicknesses for your solar array earlier in this chapter. You use the same calculation for calculating cable thicknesses for appliance cabling.

PLUGS AND SOCKETS

For 12-volt or 24-volt circuits with a current of less than 20 amps, you can use the same standard switches and light sockets as you do for normal domestic power.

However, you must not use the standard domestic plugs and sockets for attaching low-voltage devices to your low-voltage circuit. If you do, you run the risk that your low-voltage devices could accidentally be plugged into a high-voltage circuit, which could have disastrous consequences.

Instead, you have the choice of using non-standard plugs and sockets or of using the same 12-volt plugs and sockets as used in caravans and boats. These low-voltage sockets do not need to have a separate earth (ground) wire, as the negative cable should always be earthed (grounded) on a DC circuit system.

APPLIANCES

So far, I have talked a lot about 12-volt appliances, but you can buy most low-voltage appliances for either 12-volt or 24-volt and a lot of them are switchable between 12 and 24 volts.

Compared to appliances that run from grid-level voltages, you often pay more for low-voltage appliances. This is not always the case, however, and with careful shopping around, items like televisions, DVD players, radios and laptop computers need not cost any more to buy than standard versions.

Lighting

12-volt and 24-volt lighting is often chosen for off-grid solar electric systems, due to the lower power consumption of the lower-voltage lighting. You can buy low-voltage, energy-saving bulbs and strip lights, both of which provide the same quality of light as conventional lighting. LED light bulbs have vastly improved over the past five years, giving an excellent quality of light and using very little power. Filament light bulbs are also available in low-voltage forms, and although these are not very energy-efficient, they do provide an excellent quality of light.

You can buy a lot of 12-volt lighting from ordinary hardware stores. Many kitchen and bathroom lights work at low voltage and will work just as well from a 12-volt battery supply as they will from the 12-volt AC transformers typically used with this lighting. Diachronic flood lamps, halogen spot lamps, strip lamps and LED lights often run at 12 volts, giving you an excellent choice. Buying these from a hardware store rather than from a specialist solar supplier can also save a considerable amount of money.

Refrigeration

A good selection of refrigerators and freezers are available that will run from 12-volt and 24-volt power supplies. Some refrigerators will run on both low-voltage DC and grid-level AC voltage, and some can run from a bottled gas supply as well.

Unlike most other devices that you will use, refrigerators need to run all the time. This means that, although the power consumption can be quite low, the overall energy consumption is comparatively high.

There are three types of low-voltage refrigerator available:

- *Absorption fridges* are commonly found in caravans and can often use 12-volt, grid-level voltage and bottled gas to power the fridge. These are very efficient when powered by gas, but efficiency when powered on lower voltages varies considerably for different models
- *Peltier effect coolers* are not really fridges in their own right; they are portable coolers, of the type often sold in car accessory shops and powered by the 12-volt in-car accessory socket. Whilst these are cheap, most of them are not very efficient. Avoid using these for solar applications
- *Compressor fridges* use the same technology as refrigerators in the home. They are the most efficient for low-voltage operation. They are more expensive than other types but their efficiency is significantly better. Many models now consume less than 5 watts of electricity per hour

You can choose to use a standard domestic fridge for your solar electric system, running at grid-level voltages. However, they are typically not as efficient as a good 12-volt/24-volt compressor fridge. Domestic fridges also tend to have a very high starting current, which can cause problems with inverters.

A few manufacturers now produce refrigerators that are specifically designed to work with solar power. Companies such as Waeco, Sundanzer and Shoreline produce a range of refrigerators and freezers suitable for home, medical and business use.

If you wish to use a standard domestic fridge, speak to the supplier of your inverter to make sure the inverter is suitable. Many refrigerators have a very high start-up current and you may need to buy a larger inverter that can handle this sudden demand.

Microwave ovens

Standard domestic microwave ovens consume a lot more power than their rated power as their official rated power is output power, not input. You will find the input power on the power label on the back of the unit, or you will be able to measure it using a watt meter.

Typically, the input power for a microwave oven is 50% higher than its rated power.

Low-voltage microwave ovens are available, often sold for use in caravans (travel trailers) and recreational vehicles (RVs). They tend to be slightly smaller than normal domestic microwaves and have a lower power rating, so cooking times will increase, but they are much more energy-efficient.

Televisions, DVDs, computer games consoles and music

Flat screen LCD televisions and DVD players designed for 12-volt or 24-volt operation are available from boating, camping and leisure shops. These tend to be quite expensive, often costing as much as 50% more than equivalent domestic televisions and DVD players.

However, many domestic LCD televisions (with screens up to 24-inch) and DVD players often have external power supplies and many of them are rated for a 12-volt input. Some investigations at your local electrical store will allow you to identify suitable models.

If you want to use one of these, it is worth buying a 12-volt *power regulator* to connect between the television and your battery. Battery voltages can vary between 11.6 volts and 13.6 volts, which is fine for most equipment designed for 12-volt electrics, but could damage more sensitive equipment. Power regulators fix the voltage at exactly 12 volts, ensuring that this equipment cannot be damaged by small fluctuations in voltage.

Many power regulators will also allow you to run 12-volt devices from a 24-volt circuit, and are much more efficient than more traditional transformers.

Power regulators also allow you to switch from one voltage to other low voltages, if required. For example, the Sony PlayStation 3 games console uses 8.5 volts, and with a suitable power regulator you can power one very effectively from 12-volt batteries.

Power regulators can step up voltages as well as step down. A suitable power regulator can switch the voltage from a solar battery bank to an output voltage of between 1½ volts and 40 volts, depending on the specification of the regulator.

This means that many normal household items with external power supplies, such as smaller televisions, laptop computers, DVD players, music systems and computer games, to name but a few, can be connected directly to your solar power system.

Music systems

Like televisions and DVD players, many music systems have an external power supply, and a power regulator can be used in place of the external power supply to power a music system.

Alternatively, you can build your own built-in music system using in-car components. This can be very effective, both in terms of sound quality and price, with the added benefit that you can hide the speakers in the ceiling.

Using a music system with an inverter which has a modified sine wave can be problematic. Music systems designed to run at grid-level voltages expect to work on a pure sine wave system and may buzz or hum if used with a modified sine wave inverter.

Dishwashers, washing machines and tumble dryers

Dishwashers, washing machines and tumble dryers tend to be very power hungry. There are small washing machines, twin tubs and cool-air dryers available that run on low voltage, but these are really only suitable for small amounts of washing. They may be fine in a holiday home or in a small house for one person, but are not suitable for the weekly washing for a family of four.

If you need to run a washing machine from a solar electric system, you are going to need an inverter to run it. The amount of energy that washing machines consume really does vary from one model to the next. An energy-efficient model may only use 1,100 watts, whereas an older model may use almost three times this amount.

The same is true for dishwashers. Energy-efficient models may only use 500 watts, whereas older models may use nearer 2,500 watts. If you need to run a dishwasher, you will need to use an inverter.

Tumble dryers are hugely energy inefficient and should be avoided if possible. Most of them use between 2,000 and 3,000 watts of electricity and run for at least one hour per drying cycle.

There are various alternatives to tumble dryers. These range from the traditional clothes line or clothes airer to the more high-tech low-energy convection heating dryers that can dry your clothes in around half an hour with minimal amounts of power.

If you really must have a tumble dryer, you may wish to consider a bottled gas tumble dryer. These are more energy-efficient than electric tumble dryers and will not put such a strain on your solar electric system.

Air conditioning systems

Over the past couple of years, several manufacturers have been launching solar powered air conditioning and air cooling systems. Air conditioning has traditionally been very power hungry. For this reason, solar powered air conditioning has been unaffordable, as a large solar array has been required simply to run the compressors.

In response, manufacturers have developed more efficient air conditioning systems, designed to run from a DC power source. Companies such as Solar AC, Securus, Sunsource, Sedna Aire, Hitachi and LG all supply air conditioning units designed to work with solar energy.

Other manufacturers have developed evaporative air coolers that use a fraction of the power of an air conditioning unit. Whilst these air coolers do not provide the 'instant chill' factor of a full air conditioning system, by running constantly when the sun is shining, they can provide a very comfortable living and working environment at a fraction of the cost of full air conditioning.

REPUTABLE BRAND NAMES

Most solar manufacturers are not household names, and as such it is difficult for someone outside the industry to know which brands have the best reputation.

Of course, this is a subjective list and simply because a manufacturer does not appear on this list, it does not mean the brand or the product is not good.

Solar panel manufacturers and brands

Atlantis Energy, BP Solar, Canadian Solar, Clear Skies, EPV, Evergreen, Conergy, G.E. Electric, Hitachi, ICP, Kaneka, Kyocera, Mitsubishi, Power Up, REC Solar, Sanyo, Sharp, Solar World, Spectrolab, Suntech, Uni-solar.

Solar controller and inverter manufacturers and brands

Apollo Solar, Blue Sky, Enphase, Ever Solar, Exeltech, Fronius, Kaco, Magnum, Mastervolt, Morningstar, Outback, PowerFilm, PV Powered, SMA, Solectria, Sterling, Steca, SMA SunnyBoy, Victron, Xantrex.

Battery manufacturers and brands

East Penn, Chloride, Crown, EnerSys, Exide, Giant, GreenPower, Hawker, ManBatt, Newmax, Northstar, Odyssey, Optima, Panasonic, PowerKing, Tanya, Trojan, US Battery, Yuasa.

SHOPPING LIST FOR THE HOLIDAY HOME

Because our solar electric system is being installed in the garden, 10 metres (33 feet) away from the house, we have worked out that we need to run our system at 24 volts rather than 12 volts, due to the high levels of losses in the system.

I have already calculated that I need 320 watts of power from my solar array at 24 volts. To achieve this, I will need to connect at least two 12V solar panels together to create a minimum of 24V for my system, or use one larger 24V solar panel.

There are various different options available to make a 320-watt, 24-volt solar array. After checking with a number of suppliers, I have come up with the following options:

- Buy one 320Wp 24V solar panel for a total of 320 watts of power – total cost £130 ($180 US)
- Buy two 160Wp panels for a total of 320 watts of power – total cost £150 ($205 US)
- Buy four 80Wp panels for a total of 320 watts of power – total cost £120 ($165 US)
- Buy eight 40Wp panels for a total of 320 watts of power – total cost £170 ($230 US)
- Buy six 60Wp panels for a total of 360 watts of power) – total cost £170 ($230 US)

It is worth shopping around and finding the best price. Prices can vary dramatically from one supplier to another and I have seen many cases where one supplier is selling a solar panel for over twice the price it is available from elsewhere.

Depending on what configuration I buy (and where I buy it), solar panel prices for the different combinations vary between £120 ($165 US) and £400 ($570).

Based on price and convenience, I have decided to go for the cheapest option and buy four Clear Skies 80-watt polycrystalline solar panels.

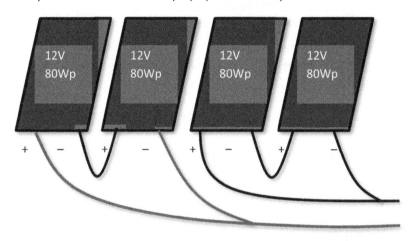

This drawing shows how I intend to wire up my solar array. I will pair two sets of panels together in series to bring the voltage up to 24 volts per pair. I then connect the pairs together in parallel to maintain the 24 volts but to increase the power of the system to a total of 320 watts

Because I am running at 24V and at a relatively low current, I have a good choice of low cost solar controllers. I decided to buy a Steca MPPT controller, which incorporates a built-in LCD display so I can see how much charge my batteries have at any one time. The cost of this controller is £120 ($155).

I calculated that I needed 181Ah of 24-volt battery storage. I have decided to go for four Northstar NSB 100FT Blue +12V, 105Ah batteries, which I will connect together in pairs to provide me 210 Ah of power at 24 volts. The cost of these batteries is £600 ($849).

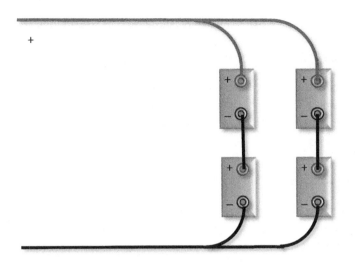

This drawing shows how I intend to wire up my batteries. I will pair two sets of batteries together in series to bring the voltage up to 24 volts per pair. I then connect the pairs together in parallel to maintain the 24 volts but to increase my storage capacity to 210 amp-hours at 24 volts

My Steca controller incorporates Ground Fault Protection, but I have decided to install a separate RCD (GFI) unit as well. It is a 'belt and braces' approach, but RCDs are extremely cheap and I feel it is worth the extra money. I still need a way of isolating the solar array manually. I choose to install three DC isolation switches: one between my controller and the solar array, one between my solar controller and my batteries and one between my controller and my distribution box. This allows me to isolate each part of my system separately, for maintenance or in case of an emergency.

For lighting, I have decided on 24-volt energy saving LED bulbs for inside use and a 24-volt halogen bulkhead light for an outdoor light. The energy saving LED bulbs look identical to grid-powered LED light bulbs and provide the same level of lighting as their grid-powered equivalents. Bulbs cost around £8/$13 each and I can use the same light switches and fittings as I would for lights powered by the grid.

I have decided to use a Shoreline RL5010 battery-powered fridge, which can run on either 12-volt or 24-volt power supplies. This has a claimed average power consumption of 9.6 watts per hour and costs £380 ($610).

For television, I have chosen a JVC 24-inch flat screen TV with built-in DVD player. The Meos TV can run on 12-volt or 24-volt power and has an average power consumption of 26 watts.

At this stage, I now know the main components I am going to be using for my holiday home. I have not gone into all the details, such as cables and configuration. We need to complete that as we plan the detailed design for our solar energy system.

IN CONCLUSION

- When choosing solar panels, buy from a reputable manufacturer if you require long life or best performance. Consider lower cost panels only for shorter term applications
- Batteries come in various types and sizes. You can calculate the optimum size of battery based on cycle life when operating on your system
- The voltage you run your system at will depend on the size of current you want to run through it. High-current systems are less efficient than low-current systems, and low- current inverters and controllers are inevitably cheaper
- Allow for future expansion in your system by buying a bigger controller and inverter than you currently need, unless you are absolutely certain your requirements are not going to change in the future
- Many appliances and devices are available in low-voltage versions as well as grid-level voltage versions. Generally, the low-voltage versions tend to be more efficient
- When wiring in 12-volt or 24-volt sockets, do not use standard domestic power sockets. If you do, you are running the risk of low-voltage devices being plugged into grid-level voltage sockets, which could have disastrous consequences

PLANNING, REGULATIONS AND APPROVALS

Depending on where you live around the world, there are different planning requirements, regulations and approvals needed for installing a solar energy system. Some countries no regulation in place; other countries have extremely tight regulations. In some countries, the regulations change between regions.

Consequently, it is impossible to provide every bit of relevant information here. Instead, we provide much of this information on *www.SolarElectricityHandbook.com*. You will also be able to find information from your local planning authority and electricity providers.

Wherever you live around the world, if you are unsure about building and electrical regulations and approvals in your area, check with an expert or ask the relevant authorities. Ignorance is never an excuse. In the case of a solar installation, the people you need to speak to are your local planning office, your buildings insurance provider and, if you are building a grid-tie system, your local electricity company. Not only will they be able to help ensure you do not fall foul of any regulation; you will often find they are a helpful and useful source of information.

NATIONAL AND INTERNATIONAL STANDARDS FOR SOLAR COMPONENTS

In the United States, Canada, Australia and across Europe, solar panels and inverters must comply with specific standards to be used in a grid-tie system. The units are tested to ensure that they conform to these standards before they are allowed on sale.

Across Europe, solar panels must be certified to IEC safety standard IEC 61730 and performance standards IEC 61215 or IEC 61646. Solar grid-tie inverters must conform to IEC 62109. Some European countries have additional certification. In Germany, grid-tie inverters must have a VDE126 certification, whilst in the United Kingdom, grid tie inverters that produce fewer than 16 amps of peak power (3.6kW) must have G83/1 certification, and larger inverters either the much more complicated G59/1 certification. Also in the United Kingdom, solar panels and inverters have to be certified by the Micro generation Certificate Scheme (MCS) to be eligible for energy export payments.

In the United States, solar panels, solar cables and inverters must have UL certification. Solar panels must conform to the UL 1703 standard. Grid-tie inverters must conform to UL 1741 and solar cabling must conform to UL 4703 or UL 854 (USE-2). If you are using batteries in your design, the batteries must conform to either UL 1989, UL 2054, UL-SU 2580 or UL-SU 1973.

In Canada, solar panels must conform to safety standard ULC/ORD-C1703-1 and design standards CAN/CSA C61215-08 or CAN/CSA C61646-2, whilst grid-tie inverters must conform to CSA C22.2 number 107.1. Batteries must conform to CAN/CSA F382-M89 (R2004). In Australia, solar panels must conform to AS/NZS5033, whilst grid-tie inverters must conform to AS4777. If you are planning a stand-alone system in a building, your system must also conform to AS4509. If you are planning a mobile system, for instance in a caravan or recreational vehicle, your system must conform to AS3001.

It is worth noting that, in all these regions, no differentiation is made between grid-tie solar and stand-alone systems for component selection. If you are building a solar energy system that is to be fitted to a building, your system must use certified components to comply with building and electrical safety regulations in these regions. If you use non-approved equipment in a grid-tie system in these countries, you will not be allowed to connect your system to the grid. You are also likely to be in contravention of building regulations and may invalidate your buildings insurance.

INSTALLATION REGULATIONS

In many countries, including the United States, Canada, Australia, New Zealand and throughout most of the European Union, you cannot work on building electrics unless you are a qualified electrician. Some countries allow you to work on electrics, but your work has to be checked and certified by a qualified electrician before commissioning. In the main, low-voltage DC circuits are excluded from this legislation, but check that this is the case. For example, Australia has stringent standards for extra-low voltage installations, including for stand alone and mobile applications.

In many countries, there are additional qualifications for electricians that allow them to install and certify solar energy systems. In most cases, it is not a legal requirement to have this additional training. However, if you wish to get access to government subsidies, feed-in tariffs or renewable energy certificates, you will almost certainly need to have your system installed, or at least checked, tested and certified, by qualified solar installation specialists. This is certainly the case in the United Kingdom and Australia. In the United States, subsidies vary from state to state, and often from county to county.

As well as electrical safety regulations, there are often regulations for the fitting of the solar panels themselves, if they are being mounted onto a building. Buffetting from wind can be an issue with solar panels. In extreme conditions, badly installed solar panels can be damaged, or can cause damage to the roofs they are fitted to. Consequently, there is legislation in many countries, including most of Europe, that states that solar panels must not be fitted within 30cm – 12 inches – of the edge of a roof.

GETTING YOUR ELECTRICITY SUPPLIER INVOLVED

If you are planning a grid-tie system, it is worth speaking to your electricity supplier involved in your project. Sometimes they have their own requirements or lists of approved equipment. They often have specialists you can speak to, who can give you extra advice and support. In some parts of the world, your electricity company will usually need to be involved while your system is being installed, replacing your current electricity meter with a specific import/export meter and carrying out the final inspection before approving your system.

Some electricity companies will only accept feed-in connections from professional solar PV installers. Almost all electricity providers insist that the installation is inspected and signed off by a certified solar installer before they will accept your connection onto the grid.

IN CONCLUSION

- There are different rules and regulations for installing solar power around the world
- There are many standards, covering both the actual physical hardware and how it is installed
- You must comply with the building regulations and electrical regulations in force in your region
- You will be able to find help by talking to your local planning office and your electricity provider. You will also need to talk to your building insurance company

DETAILED DESIGN

By now, you know what components you are going to use for your solar project. The next step is to work on your detailed design and create a picture of what you want to build. Even for simple projects, it makes sense to draw up a diagram before installation.

There are many benefits of drawing a wiring diagram:

- It ensures that nothing has been overlooked
- It will assist in the cable sizing process
- It helps ensure nothing gets forgotten in the installation (especially where there is a group of people working together on site)
- It provides useful documentation for maintaining the system in the future

The wiring diagram will be different for each installation and will vary depending on what components are used. Read the product documentation for each component for information on how it must be wired.

If you have not yet chosen your exact components at this stage, draw a general diagram but make sure that you flesh this out into a detailed document before the installation goes ahead.

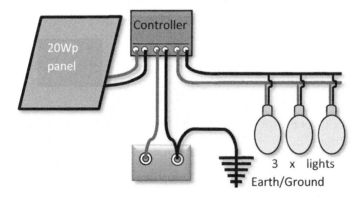

A sample wiring diagram for a simple stand-alone lighting system

When drawing up your wiring diagrams, you will need to remember the following:

SAFETY IS DESIGNED IN

It is easy to forget that solar energy can be dangerous. We are working with electricity and whilst any individual component may only be low-voltage, some of the currents involved can be quite significant. Connecting multiple solar panels or batteries together in series very quickly creates a high voltage. Safety must be taken into account during the detailed design phase of the project, as well as during installation.

When designing the system, ask yourself this question:

"What's the worst that can happen?"

Solar energy systems are relatively straightforward and the design of all the components you will use will keep risks to an absolute minimum. Nevertheless, there are potential risks. If you are aware of these risks, you can take steps to eradicate them in your design.

What is the worst that can happen with a solar installation?

With solar energy, we will be working in a few risk areas. Namely, DC electrics from the solar array, high currents from batteries, AC electrics if you are using an inverter, and high temperatures from the solar panels themselves. You may also be working at height if installing panels on a roof.

Each of these risk areas can pose problems, both in isolation and when combined. It is worth considering these risks to ensure that you can exclude as many of them as possible.

Grounding your electrics

Except for a very small system, such as rigging up a light in a shed, a solar energy system should always be earthed (grounded). This means running a wire from a negative terminal to an earthing rod (known as a *grounding rod* in North America) that is rammed into the ground.

An earthing rod (grounding rod) is a 1m (3 foot) long metal pole, typically made of copper. They are available from all electrical wholesalers and builders' merchants.

Connections to a ground prevent build-up of static electricity and can help prevent contact with high voltages if the circuit gets damaged.

You must always include a ground connection from the solar array itself. Whilst it is optional for very small solar arrays generating less than 100 watts, it is always a good idea to include a ground from a solar array as this will reduce the risk of a short-circuit fault resulting in a fire. You must also earth the battery bank, as they can deliver very high currents.

If you are using both AC electrics and DC electrics in your system, you must always have a separate ground for each system.

Grounding a system where you cannot connect to the ground

There may be instances where you are building a system where no connection to the ground is possible. For instance, a portable solar charging unit that can be carried anywhere, or a solar powered boat.

Typically, these designs are very small, using only DC electrics and running only a few amps of current. If your solar array is less than 100 watts, your system runs at 12 volts and you are drawing less than 10 amps of current, you are unlikely to need a common earth for all your components.

For larger systems, a *ground plane* is often used. A ground plane is a high-capacity cable connected to the negative pole on the battery, to which every other component requiring a ground is also connected. A thick, heavy-duty battery interconnection cable is often used as a ground plane cable, with thinner wires connecting to this ground plane cable from every other component requiring an earth.

As an alternative to a high-capacity cable, depending on what you are installing your solar system on, you can use a metal frame as the common ground for your system. In standard car electrics, for example, the ground plane is the car body itself.

DC Electrics

Direct current electricity typically runs at relatively low voltages. We are all familiar with AA batteries and low-voltage transformers used for charging up devices such as mobile phones. We know that if we touch the positive and negative nodes on an AA battery we are not going to electrocute ourselves.

However, direct current electricity can be extremely dangerous, even at comparatively low voltages. A small number of people are killed every year by licking 9-volt batteries, because of the electric jolt they receive. Scale that up to an industrial grade heavy-duty 12-volt traction battery, capable of delivering over 1,000 amps of current, or a solar array capable of producing hundreds of volts on an open circuit, and it is easy to see that there is a real risk involved with DC electrics.

If you are electrocuted with AC power, the alternating current means that whilst the shock can be fatal, the most likely outcome is that you will be thrown back and let go. If you are electrocuted with DC power, there is a constant charge running through you. This means you cannot let go. If you are electrocuted with very high current DC, the injury is more likely to be fatal than a similar shock with AC power.

Because of the low current from a single solar panel, you are unlikely to notice any jolt if you short-circuit the panel and your fingers get in the way. However, wire up multiple solar panels together and it is a different story. Four solar panels connected in series produce a nominal 48 volts. The peak voltage is nearer 80–100 volts. At this level, a shock could prove fatal for a young child or an elderly person.

The current thinking with grid-tie solar systems is to connect many solar panels together in series, creating a very high-voltage DC circuit. Whilst there are some (small) efficiency benefits of running the system at very high voltage, there are risks as well, both during the installation and the ongoing maintenance of the system.

There are issues with the 12-volt batteries too. Industrial grade, heavy-duty batteries can easily deliver a charge of 1,000 amps for a short period. Short out a battery with a spanner and it will be red hot in just a few seconds. In fact, the current delivery is so great it is possible to weld metal using a single 12-volt battery.

The big risk with DC electrics is electrocuting yourself (or somebody else) or causing a short circuit, which in turn could cause a fire. Solar panels generate electricity all the time, often including a small current at night, and cannot simply be switched off. Therefore, there need to be manual DC circuit breakers (also called isolation switches) to isolate the solar panels from the rest of the circuit, plus a good ground and a ground fault protection system to automatically switch off the system should a short circuit occur.

If your system is running at a high voltage, you may want to consider multiple DC circuit breakers/ isolation switches between individual solar panels. This means that, as well as shutting off the overall circuit, you can reduce the voltage of the solar array down to that of a single panel or a small group of panels. This can be of benefit when maintaining the solar array, or in the case of an emergency.

A short circuit in a solar array can happen for many reasons. Sometimes it is because of a mistake during installation, but it can also occur because of general wear and tear (especially with installations where the tilt of the solar panels is adjusted regularly) or due to animal damage such as bird mess corroding cables or junction boxes, or a fox chewing through a cable.

Short circuits can also occur where you are using unsuitable cabling. Solar interconnection cabling is resistant to UV rays and high temperatures, and the shielding is usually reinforced to reduce the risk of animal damage. Always use solar interconnection cabling for wiring your array and for the cabling between the array and your solar controller or inverter.

When a short circuit does occur, there is often not a complete loss of power. Instead, power generation drops as resistance builds up. There is a build-up of heat at the point of failure. If you have a ground fault protection system such as an RCD or GFI in place, the system should switch itself off automatically at this point, before any further damage is caused.

If you do not have a ground fault protection system in place, the heat build-up can become quite intense, in some cases as high as several hundred degrees. There have been documented instances where this heat build-up has started a fire.

If a fire does break out, you need to be able to isolate the system as quickly as possible. A solar array cannot be switched off as it always generates power whenever there is light. Consequently, there have been cases where the fire brigade has not been able to put out a fire generated by a fault in a solar array because there has been no way of switching it off. Isolating the solar array quickly using a DC circuit breaker resolves this problem.

Remember that even if you isolate the solar array, you are still generating power within the solar array. If you have many solar panels, the voltage and the current can still be quite considerable. The ability to shut down the array by fitting DC circuit breakers or contactors within the array can significantly reduce this power, rendering the system far safer if there is an emergency.

AC electrics

AC electrical safety is the same as household electrical safety. It is high-voltage and in most countries you are not allowed to work with it unless you are suitably qualified.

You will need to install two AC isolation switches: one switch between the inverter and the distribution panel to isolate the solar system completely, and one switch between your grid-feed and your distribution panel to isolate your system from the grid if you are running a grid-tie system.

If you are planning a grid-tie installation, you will need to speak with your electricity supplier, as there will often be additional requirements that you will need to incorporate. Your inverter will need to be a specific grid-tie system that switches off in the case of a grid power cut. This ensures that power is not fed back into the grid from your solar system in the case of a power failure, which could otherwise prove fatal for an engineer working on restoring power.

High temperatures

We have already touched on the risk of high temperatures with a solar array. Solar panels are black and face the sun, consequently they can therefore get very hot on a bright day. Certainly hot enough to fry an egg and definitely hot enough to burn skin.

So make sure your solar array is installed in a place where it cannot be touched by curious children. If the solar panels are close to the ground, make sure there is some protection to keep people away from it.

The high temperatures become more of a problem if there is a fault within the solar array or with the wires running between solar panels. If a cable or a solar panel becomes damaged, there can be significant heat build-up. As already mentioned, this heat build-up can lead to a fire.

A *residual current device* (RCD), otherwise known as a *ground fault interrupter* (GFI) should avert this problem, allowing you to investigate the issue before significant damage occurs. However, manual DC circuit breakers should also be installed to override the system in case of an emergency.

Think safety

That is the end of the safety lecture for now. I will touch on safety again when we come to installation, but for now, please remember that safety does not happen by accident. Consider the safety aspects when you are designing your system and you will end up with a safe system. The additional cost of a few AC and DC circuit breakers, an earthing rod/ grounding rod, an RCD/ GFI and getting the right cables is not going to break the bank. It is money well worth spending.

SOLAR ARRAY DESIGN

All solar panels in an array must face in the same direction. This ensures that each cell receives the same amount of light, which is important for optimum power production.

Sometimes, you may want to install solar panels in different locations, such as on two different pitches of roof. In this instance, you need to keep the two banks of solar panels separate, running them as two separate arrays, either by feeding them into an inverter or controller that can handle more than one solar input, or by feeding them into two separate inverters or controllers.

If you wish to mix and match different sizes of solar panel, you will also need to set these up in separate arrays and wire these separately, either using an inverter or controller that can handle more than one solar input, or using two separate inverters or controllers.

If you are designing a grid-tie system, where you are considering different sizes or orientations of solar panels, you should seriously consider a micro-inverter system where each solar panel has its own inverter.

Solar array design – stand-alone systems

If you have more than one solar panel and you are running your solar electric system at 12 volts, then you will need to wire your panels together in parallel to increase your capacity without increasing the overall voltage, or use a solar charge controller that can accept a much higher input voltage and drop the voltage to match the battery.

If you are running your battery pack at higher voltages, you either need to buy higher-voltage solar panels, or you will need more than one solar panel, wired in series to increase the voltage of the solar panels to the voltage of your overall system:

- For a 24-volt battery system, you have the choice of using 24-volt solar panels, or two 12-volt solar panels connected in series

- For a 48-volt battery system, you can use one 48-volt solar panel, two 24-volt solar panels connected in series, or four 12-volt solar panels connected in series

Once you have reached the voltage that you want, you can then run the panels both in series and in parallel, connecting strings of panels together in series to reach your desired voltage, and then connecting multiple strings together in parallel to increase your capacity:

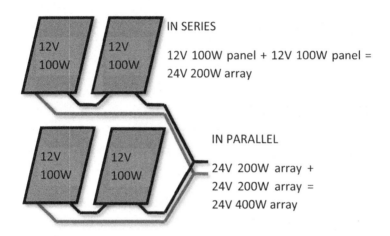

A sample diagram of a 24-volt array where two sets of two 12-volt solar panels are connected in series to create a 24-volt array and the two arrays are then connected in parallel to create a more powerful 24-volt array.

Solar array design – grid-tie systems with micro-inverters

If you are designing a grid-tie system and using micro-inverters, your design is extremely simple. Each solar panel becomes a self-sufficient solar energy system, each feeding power into its own micro-inverter. The micro-inverters convert the energy to AC and feed it into the main AC circuit.

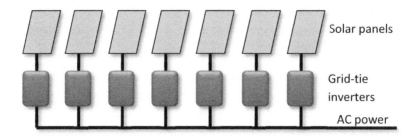

Solar array design – grid-tie systems with a single inverter

If you are designing a grid-tie system with a single inverter, you will typically be connecting all your solar panels in series and then feeding this high-voltage DC power into an inverter.

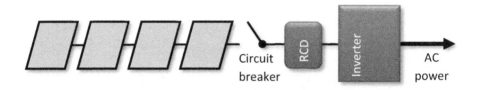

A simplified block diagram of a typical grid-tie system using a single inverter. The residual current device (RCD) provides ground fault protection. In the United States, an RCD is known as a ground fault interrupter (GFI)

Because of the very high DC voltages involved, additional safeguards are necessary. The solar array must always be grounded, there must be a DC circuit breaker (also known as an isolation switch) installed between the solar array and the inverter and there must be a DC residual current device/ground fault interrupter installed to shut down the solar array in the case of a short circuit.

In the diagram below, there are sixteen 250Wp solar panels connected in series. This is a fairly typical installation for a residential property, creating a 4kWp system. Assuming each solar panel produces a 24-volt output, this system will run at a nominal 384 volts, with a peak power in the region of 640 volts and an open circuit voltage of up to 832 volts.

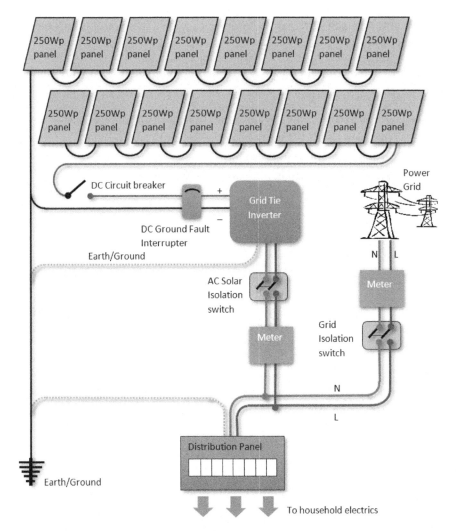

Above: A sample block diagram for a grid-tie system

This system would be legal in Europe, but breaks building regulations in the United States which disallows any system that has the potential to run at over 600V. Consequently, this system would more typically be installed with the solar array split into two smaller strings, and running with a grid-tie inverter capable of accepting two separate solar inputs.

In Europe, voltages of up to 1,000V are allowed. However, the dangers of such high voltages means that even in Europe, it is common practice to split solar arrays to reduce voltages down to under 500-600V where possible. In general, this means that you will not want to connect more than twenty 12-volt solar panels or ten 24-volt solar panels in series, to ensure that you stay well below this level.

If you are running close to the 600V limit in the United States or 1,000V limit in Europe, there are three options:

- Install a multi-string system, either using an inverter that handles more than one solar feed, or by using two separate inverters
- Install a micro-inverter system
- Wire your solar panels in a parallel/series hybrid

For more information about open circuit voltages, refer back to page 120. To recap on multi-string systems, refer back to the section on Multiple strings on page 124.

BATTERIES

Batteries are wired in a similar way to your solar array. You can wire up multiple 12-volt batteries in parallel to build a 12-volt system with higher energy capacity, or you can wire multiple batteries in series to build a higher-voltage system.

When wiring batteries together in parallel, it is important to wire them up so that you take the positive connection off the first battery in the bank and the negative connection off the last battery in the bank.

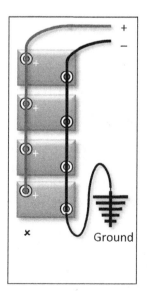
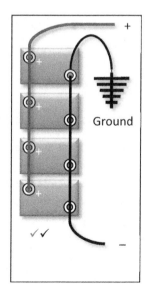

*How to wire batteries in parallel: the diagram on the left, where power feed for both positive and negative is taken off the first battery in the bank, shows how **not** to do it – it will lead to poor battery performance and premature battery failure. The diagram on the right, where the positive feed is taken off the first battery in the bank and the negative feed is taken off the last battery in the bank, is **correct** and will lead to a more balanced system with a significantly longer life.*

This ensures equal energy drain and charging across the entire battery bank. If you use the same battery in the bank for negative and positive connections to the controller and inverter, you drain this first battery faster than the rest of the batteries in the bank. The first battery also gets the biggest recharge from the solar array.

This shortens the life of the battery and means all the batteries in the bank end up out of balance. Other batteries in the bank never get fully charged by the solar array, as the first battery will report being fully charged first and the controller will then switch the power off rather than continuing to charge the rest of the batteries in the bank. The result is that the batteries end up with a shorter lifespan.

As discussed on the chapter on site surveys, the location of your batteries needs to fit the following criteria:

- Water- and weather-proof
- Not affected by direct sunlight
- Insulated to protect against extremes of temperature
- Facilities to ventilate gases (specifically with regards to lead acid batteries)
- Protected from sources of ignition
- Because of the extremely high potential currents involved with lead acid batteries, the batteries must be in a secure area away from children and pets.

You need to ensure that your batteries are accessible for regular checks and maintenance. Many deep-cycle batteries require watering several times each year and connections must be checked regularly to ensure they have not corroded.

If you are installing batteries inside, batteries are often mounted on heavy-duty racking, which is then made secure using an open-mesh cage, or in a dedicated battery box.

If you are installing your batteries in an area that can get very cold or very hot, you should also insulate your batteries. Most batteries perform at their best at between 15–28°C (59–77°F). Very few batteries will work at all below 0°C (32°F) or above 45°C (113°F). Extreme temperatures adversely affect the performance of batteries, so if your batteries are likely to be in an area where the temperature drops below 8°C (46°F) or rise above 30°C (86°F), you should consider providing insulation. If the temperature is likely to drop below freezing, you must provide it.

Both the process of charging and discharging a battery generates a certain amount of heat, which means that for most applications where the batteries are in use for most of the time, extreme cold is not such a big issue. Ensuring batteries are not placed on a cold concrete floor (a wood base is much warmer) can make a big difference to winter performance. Likewise, if you are designing a system that is rarely used in the depths of winter, ensuring the batteries remain above 0°C (32°F) is not such a big issue as the low temperatures are unlikely to cause permanent damage to the batteries.

Excessive heat in the height of summer is more of an issue. For this reason, placing batteries in a loft or roof-space is not always a good idea as these areas can easily reach 50–60°C (122–140°F), even in relatively cool climates, unless air extraction is built into the roof space to keep temperatures under control. Installing batteries into a dedicated battery box, or into a garage, is a far better solution.

Creating an insulated environment for batteries helps with both extremes of temperature. You can use polystyrene (Styrofoam) sheets underneath and around the sides of the batteries to keep them insulated. Alternatively, foil-backed bubble-wrap insulation (available from any DIY store in the insulation section) is even easier to use and has the benefit that it does not disintegrate if you ever get battery acid splashed on it.

Never insulate the top of the batteries, as this will stop them from venting properly and may cause shorts in the batteries if the insulating material you use is conductive.

If you are using batteries that can vent hydrogen, you need to create a design that allows the hydrogen to escape. Hydrogen rises rapidly from the batteries and can easily get trapped in a roof-space where it can become dangerous if it can build up. As mentioned in the section on Venting Hydrogen from Lead Acid batteries on page 150, you need to ensure that hydrogen can vent safely out of a building, and fit hydrogen alarms in a roof space above a battery pack if fitting batteries inside a building.

Controller

A controller will have connections to the solar array, to the battery bank and to DC loads. Although controllers tend not to have the same heat problems as inverters, they can get warm in use. Make sure they are installed in an area with good ventilation around them and in a location where they can be easily checked.

Inverter

Where an inverter is used in a stand-alone or grid fallback system, it is connected directly to the battery bank and not through the controller.

Make sure that you design your system so that the inverter is in a well-ventilated area, as they can generate a fair amount of heat. Most inverters are around 97% efficient, with the extra 3% lost in heat. Whilst this may not sound a lot, it can soon add up if the inverter is boxed in: a 4kW inverter can easily generate 120W of heat per hour when running at peak performance in the summer. Over the period of a few hours, this can easily be enough to cause the inverter to overheat and shut down.

Consider the weight of the inverter. Also ensure that it is installed in a location where it can easily be checked. Most inverters incorporate a small display that allows you to monitor system performance.

Devices

Devices are connected to the inverter if they require grid-level voltage, or to the controller if they are low-voltage DC devices. They are never connected directly to the solar array or the batteries.

SPECIFICS FOR A GRID FALLBACK SYSTEM

Because a grid fallback system usually does not connect your solar energy system to the grid, you are less restricted as to the components you can use.

You must still adhere to basic wiring legislation for your country. In some countries (such as the United Kingdom, for instance) this can mean having the final connection into your building electricity supply installed by a fully qualified electrician, but this is significantly cheaper than having a grid-tie system installed and inspected.

The design for a grid fallback system is very similar to a stand-alone solar system, i.e. solar panels, solar controller and batteries. The only difference is what happens after the batteries. The advantage of a grid fallback system is that it can work in three ways. It can provide power for an entire building, it can provide power for specific circuits within a building or it can provide power for a single circuit within a building.

More information and a sample circuit diagram for grid fallback configurations are shown in Appendix B.

CIRCUIT PROTECTION

Circuit protection is required in any system to ensure the system shuts down safely in the event of a short circuit. It is as valid on low-voltage systems as it is on high-voltage systems.

A low-voltage system can cause major problems simply because of the huge current that a 12-volt battery can generate. A deep-cycle lead acid battery can produce more than 1,000 amps in a short burst, which can easily cause a severe shock and even death or serious injury in some cases.

In the case of a short circuit, your wiring will get extremely hot and start melting within seconds unless suitable protection has been fitted. This can cause fire or burns, and necessary protection should be fitted to ensure that no damage to the system occurs as a result of an accidental short circuit.

Earthing (grounding)

In all systems, the negative terminal on the battery should be adequately earthed (referred to as *grounded* in North America). If there is no suitable earth available, a grounding rod or ground plane should be installed.

DC circuit protection

For very small systems generating less than 100 watts of power, the fuse built into the controller will normally be sufficient for basic circuit protection, although an external fuse is generally recommended. In a larger system, where feed for some DC devices does not go through a controller, a fuse should be incorporated on the battery positive terminal.

Where you fit a fuse to the battery, you must ensure that all current from the battery must pass through that terminal. In DC systems with multiple circuits, it is advisable to fit fuses to each of these circuits. If you are using 12 volts or 24 volts, you can use the same fuses and circuit breakers as you would for normal domestic power circuits. For higher-voltage DC systems, you must use specialist DC fuses.

When connecting devices to your DC circuits, you do not need to include a separate earth (ground) for each device, as the negative is already earthed at the batteries.

Fit an isolation switch (DC disconnect switch) between your solar array and your inverter or controller. Fit a second isolation switch between your batteries and your controller and inverter. Unless your controller or inverter already incorporates one, you should fit a DC residual current device/ ground fault interrupter between your solar array and your controller or inverter.

AC circuit protection

AC circuits should be fed through a distribution panel (otherwise known as a consumer unit). This distribution panel should be earthed (grounded) and should incorporate an earth leakage trip with a residual current device (RCD), otherwise known as a ground fault interrupter.

As you will have earthed your DC components, you must use a separate earth (ground) for AC circuits. You must also install an AC disconnect switch (isolation switch) between your inverter and your distribution panel. In the case of a grid-tie system, this is normally a legal requirement, but it is good practice anyway.

The wiring in the building should follow normal wiring practices. You should use a qualified electrician for installing and signing off all grid-level voltage work.

CABLE SIZING AND SELECTION

Once you have your wiring diagram, it is worth making notes on cable lengths for each part of the diagram, and making notes on what cables you will use for each part of the installation.

Sizing your cables

This section is repeated from the previous chapter. I make no apologies for this, as cable sizing is one of the biggest mistakes that people make when installing a solar electric system.

Low-voltage systems lose a significant amount of power through cabling. This is because currents (amps) are higher to make up for the lack of voltage. Ohms law tells us that the power lost through the cable is proportional to the square of the current: the higher the current, the greater the energy loss. To overcome this energy loss, we must either increase the voltage or use thicker cables.

Wherever you are using low-voltage cabling (from the solar array to the controller, and to all low-voltage DC equipment) you need to ensure you are using the correct size of cable. If the cable size is too small, you will get a significant voltage drop that can cause your system to fail. Your cables can also get hot, and in extreme cases, could melt or start a fire. Never use smaller cable, as this could cause some of your devices not to work properly.

Protecting cable runs

When planning cable layouts, you need to ensure they are protected from unwanted attention from animals and children and from possible vandalism.

Rats and foxes chewing through cable insulation can be a big problem in some installations. This can be resolved by using rodent protected cabling. Using conduit is often a good idea as well, especially if you can use steel conduit or thin-wall electrical metallic tubing (EMT) to protect cables.

Designing your system to keep your cable runs as short as possible

If you have multiple devices running in different physical areas, you can have multiple cable runs running in parallel to keep the cable runs as short as possible, rather than extending the length of one cable to run across multiple areas.

By doing this, you achieve two things. Firstly, you are reducing the overall length of each cable and secondly, you are splitting the load between more than one circuit. The benefit of doing this is that you can reduce the thickness of each cable required, which can make installation easier.

If you are doing this in a house, you can use a distribution panel (otherwise known as a consumer unit) for creating each circuit.

In the holiday home, for instance, it would make sense to run the upstairs lighting on a different circuit to the downstairs lighting. Likewise, it would make sense to run separate circuits for powering appliances upstairs and downstairs.

In the case of the holiday home, by increasing the number of circuits it becomes possible to use standard 2.5mm domestic 'twin and earth' cable for wiring the house, rather than more specialist cables. Not only does this simplify the installation, it keeps costs down.

Selecting solar cable

A common fault with poorly-designed or poorly-installed solar energy systems is under-performance where there is no clear source for the problem. In particular, this tends to occur around two to three years after the system was first installed.

The source of the problem is often either bird droppings or UV damage on cables leading from the solar panels to the inverters. This is usually caused by not using solar interconnection cables, which have a much tougher insulation that is UV protected, designed to withstand high temperatures and can withstand acidic bird droppings.

It is vital that you use specific solar interconnection cable to connect your solar panels together and for linking your solar panels to your inverter or controller. If you are not sure, look for cable that conforms to the UL 4703 or UL 854 (USE-2) specification for PV cabling. This is available from all solar equipment suppliers.

Controller cable

When calculating the thickness of cable to go between the controller and the battery, you need to take the current flow into the battery from the solar array as well as the flow out of it (peak flow into the battery is normally much higher than flow out).

Battery interconnection cables

You can buy battery interconnection cables with the correct battery terminal connectors from your battery supplier. Because the flow of current between batteries can be very significant indeed, I tend to use the thickest interconnection cables I can buy for connection between batteries.

SOME SAMPLE WIRING DIAGRAMS

As ever, a picture can be worth a thousand words, so here are some basic designs and diagrams to help give you a clearer understanding of how you connect a solar electric system together.

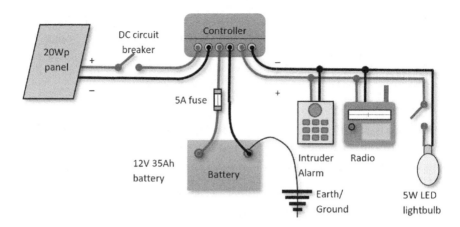

Above: A simple solar installation: a light with light switch, a small radio and a simple intruder alarm – perfect for a beach hut, garden shed or a small lock-up garage.

Because it is a small system, you may choose not to fit an isolation switch. Because the system is very small, I have decided only to fit a fuse between the controller and the battery

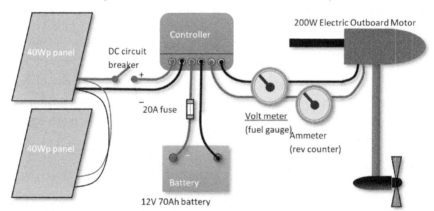

Above: A solar powered river boat. Electric boats are terrific on the river – they're virtually silent allowing you to get close to nature and all you hear is the lapping of the water; the only downside is recharging the batteries.

Here, solar panels are used to recharge the batteries, charging them up during the week to provide all the power required for a weekend messing about on the river.

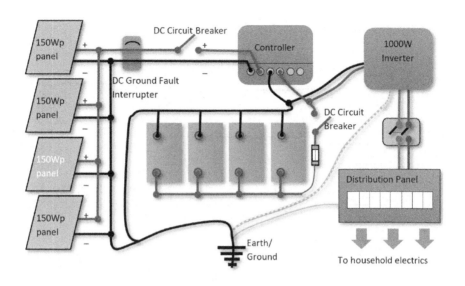

The holiday home wiring diagram

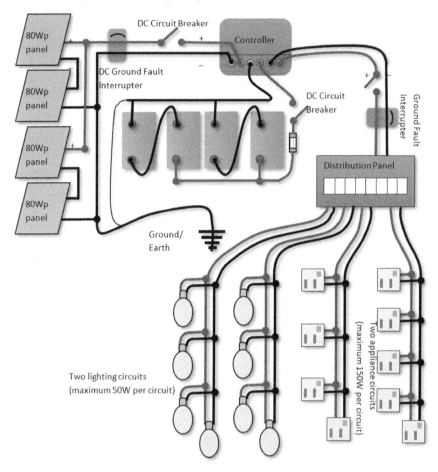

THE NEXT STEP

Once you have your wiring diagram, it is time to start adding cables, battery terminal clamps, fuses, isolation switches, earthing rods (referred to as *grounding rods* in North America) and, in this case, a distribution panel (otherwise known as a consumer unit) to your shopping list. It can help to add more detail to your wiring diagram as well, noting the locations of appliances and sockets, and the lengths of cables at each point.

SOLAR FRAME MOUNTING

As discussed in the section on Solar panel mountings, starting on page 106, there are off-the-shelf solar array frames available, and your solar panel supplier will be able to advise you on your best solution.

Sometimes, these are not suitable for your project. In this case, you will either fabricate something yourself (angle iron is a useful material for this job) or get a bespoke mounting made specifically for you.

Solar panels in themselves are not heavy, but you do need to consider the effect of wind loadings on your mounting structure. If the wind can blow underneath the solar array it will generate a 'lift', attempting to pull the array up off the framework. However, a gap beneath the solar array is useful to ensure the array itself does not get too hot. This is especially important in warm climates, where the efficiency of the solar panels themselves drops as they get hotter.

Making sure the mounting is strong enough is especially important as the solar array itself is normally mounted at an optimal angle to capture the noonday sun. This often means that, even if you are installing your solar array onto an existing roof, you may want to install the solar panels at a slightly different angle to the roof itself to get the best performance out of your system.

It is therefore imperative that your solar array mounting frame is strong enough to survive 20 years plus in a harsh environment and can be securely mounted.

If you are mounting your solar array on a roof, you must be certain that your roof is strong enough to take this. If you are not certain about this, ask a builder, structural surveyor, or architect to assess your roof.

If you are planning to mount your solar array on a pole or on a ground-mounted frame, you will need to make plans for some good strong foundations. Hammering some tent pegs into the ground to hold a ground-mounted frame will not last five minutes in a strong wind, and a pole will quickly blow down if you only use a bucket of cement to hold it in place.

You should build a good foundation consisting of a strong concrete base on a compacted hardcore sub-base to hold a ground-mounted frame, and the frame itself should be anchored using suitable ground anchors, bolted using 25cm–30cm (10"–12") bolts.

For a pole, follow the advice given by the manufacturers. Typically, they need to be set in a concrete foundation that is at least 3 feet (1m) deep, and quite often significantly more.

To mount your solar panels onto your frame, make sure you use high-tensile bolts and self-locking nuts to prevent loosening due to wind vibration.

If your solar array is going to be easily accessible, you may wish to consider an adjustable solar mounting system so you can adjust the angle of tilt throughout the year. You can then increase the tilt during the winter to capture more winter sun and decrease the tilt during the spring and summer to improve performance during those seasons.

For the holiday home project, the solar array is to be fitted to a specially constructed garden store with an angled roof. The benefit of this approach is that we can build the store at the optimum position to capture the sun. We can also install the batteries and solar controller very close to the solar array. In other words, we are creating an 'all in one' power station.

There are many regional shed and garden building manufacturers who will build a garden store like this to your specification. A good quality store, so long as it is treated every 2–3 years, will easily last 25–30 years.

If you go this route, make sure your chosen manufacturer knows what you are planning to use it for. You need to specify the following things:

- The angle of the roof must be accurate to have the solar panels in their optimum position
- The roof itself must be reinforced to be able to take the additional weight and potential wind lift of the solar array
- The floor of the garden store (where the batteries are stored) must be made of wood. Batteries do not work well on a concrete base in winter
- There must be ventilation built into the store to allow the hydrogen gas generated by the batteries to disperse safely through the top of the roof
- The door to the garden store itself should be large enough for you to easily install, check and maintain the batteries
- You should consider insulating the floor, walls and ceiling in the garden store, either using polystyrene (Styrofoam) sheets or loft insulation. This will help keep the batteries from getting too cold in winter or too hot in summer

A garden store will still require a solid concrete foundation. Consideration of rainwater runoff is also important, to ensure the garden store does not end up standing in a pool of water.

Roof mounting solar PV

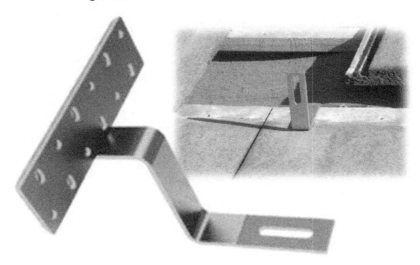

Roof-mounting kits are available from solar panel suppliers. These are made up of roof hooks that fit to the roof trusses underneath the tiles, to which mounting rails are attached. Typically two mounting rails are required for each row of solar panels. Clamps are then used to clamp the solar panels to the mounting rails.

Roof Hooks

To install solar panels onto an existing tiled roof, individual tiles are slid out and a roof hook is fitted beneath the tile, mounting directly onto the underlying roof truss. Once mounted, the original tile can be slid back into position. Depending on the type of tile and the roof hook design, it may be necessary to chisel out part of the underside of the tile to achieve a good fit.

Mounting Rails

Mounting rails are then fitted horizontally across the roof, bolted to the roof hooks using locking nuts to ensure they cannot work loose over time. Two rows of mounting rails are used to hold each row of solar panels.

Clamps

Clamps are then slid into place along the mounting rails. These are used to clamp the edges of each solar panel into position.

Finally, the first solar panel can be fitted into position. Four clamps are used to hold a solar panel into position. However, each clamp can hold two solar panels down when the panels are next to each other.

PLANNING THE INSTALLATION

By now, you should have a complete shopping list for all the components you need. You should know where everything is to be positioned and what you need to to proceed.

Before placing any equipment orders, go back to your site and check everything one last time. Make sure that where you planned to site your array, controller, batteries and so on is still suitable and that you have not overlooked anything.

Once you are entirely satisfied that everything is right, place your orders for your equipment.

Bear in mind that some specialist equipment is often only built to order and may not be available straight away. If you require bespoke items such as solar mounting frames, or, as in the case of the holiday home, a complete garden store made up for mounting the solar panels and holding the batteries and controller, take into account that this could take a few weeks to be built for you.

IN CONCLUSION

- The detailed design ensures you have not overlooked any area of the design
- Consider the safety aspects of your system in your design. At each stage, ask yourself *"What is the worst that can happen?"* and then design around the problems
- The wiring diagram helps you envisage how the installation will work
- You need to keep cable runs as short as practically possible. You can do this by running several cables in parallel, either directly from the controller, through a junction box or through a distribution panel
- Splitting the cables into parallel circuits also means you reduce the current load on each circuit, thereby reducing resistance and improving the efficiency of your system
- If you are using an inverter to run at grid-level voltages, a qualified electrician is required to handle the electrical installation. However, your wiring diagram will help your electrician to envisage how your solar electric system should work
- You need to design your battery storage area to ensure your batteries can perform to the best of their ability

INSTALLATION

Congratulations on getting this far. If you are doing this for real, you will now have a garage or garden shed full of solar panels, batteries, cables, controllers, isolation switches, mounting brackets and RCDs. The planning stage is over and the fun is about to begin.

Before you get your screwdriver and drill out, there are just a few housekeeping items to get out of the way first...

SAFETY

First, there are a few safety points to be aware of. Some of these may not be relevant to you, but read them all first, just to make sure. Remember, you are working with electricity, dangerous chemicals and heavy but fragile objects. It is better to be safe than sorry.

Your First Aid kit

You will need a good First Aid kit on hand, including some items that you will not normally have in a regular First Aid kit. Most specifically, you will need an eye-wash and a wash kit or gel that can be applied to skin in case of contact with battery acid.

Chemical clean-up kit

If you will be working with lead acid batteries that contain chemicals that are hazardous to health. You will require the following:

- A chemical clean-up kit suitable for cleaning up battery acids in the case of a spill
- A supply of strong polythene bags
- A good supply of rags/ disposable wipes to mop up any battery spillages

Chemical clean-up kits and chemical First Aid kits are available from most battery wholesalers and industrial tool suppliers. They only cost a few pounds. You probably will not need them but, if nothing else, they buy you peace of mind.

Considering the general public

If you are working in an area where the general public has access, you should use barriers or fencing, and signage to cordon off the area. Clear diversion signage should explain an alternative route.

In this scenario, I would recommend employing a professional team of builders to carry out the installation work on your behalf. They will already understand the implications of working in a public area and the relevant Health and Safety regulations.

Even if you do not have to consider the general public, you should still consider the people around you. Children love to get involved with these sorts of projects and there really can be some safety issues involved. Keep children out of the way, and let anyone in the vicinity know that you are working with high voltages and to keep away.

Working at height

You are very likely to be working at height and quite possibly crawling around on slanted rooftops.

Make sure you are using suitable climbing equipment (ladders, crawler boards, safety harnesses, scaffolding). You can hire anything that you do not have at reasonable prices.

If you have any concerns about working at heights, or if you are working beyond your area of competence at any time, remember there is no shame in hiring a professional. A professional builder can fit a solar array to a roof in 2–3 hours. This is typically less than half the time it takes an amateur DIY enthusiast.

Handling

Batteries, large inverters and solar arrays can be heavy. Solar panels themselves may not be heavy in their own right, but when several of them are mounted on a frame and then lifted they are heavy, bulky and fragile.

Moving and installing much of this equipment is a two-person job as a minimum. More people can be useful when lifting a solar array into position.

Working with batteries

Lead acid batteries are extremely heavy, in some cases weighing as much as an adult. Use proper lifting gear to move them, and look after your back.

Heavier batteries quite often have hoops in the top case. To lift a battery, I tend to use a piece of rope threaded through these hoops to create a carrying handle. This means I can carry a battery close to the ground and reduce the need to bend over to lift it.

Lead acid batteries contain acid. Unless they are gel batteries, the acid is in liquid form. It is extremely corrosive and extremely dangerous to health. Splashes of liquid from the batteries can cause severe chemical burns and must be dealt with immediately.

When working with lead acid batteries, stay safe:

- ALWAYS wear protective clothing, including overalls, eye protection (either protective glasses or a full-face shield) and protective gloves. I would also advise you to wear steel toe-capped shoes
- Keep batteries upright at all times
- Do not drop a battery. If you do, the likelihood is that the battery has been damaged. In the worst-case scenario, the casing could be cracked or broken
- If you drop a battery, place it immediately in a spill tray (a heavy-duty deep greenhouse watering tray can be used if necessary) and check for damage and leaks
- If you have a damaged battery, both the battery and the spill tray must be double-bagged in sealed polythene bags and marked as hazardous waste
- If you have a spillage from a battery, mop up the spillage immediately using rags or disposable wipes. Place these rags in a polythene bag, seal it and mark it as hazardous waste
- If any spillage from a battery comes into contact with clothing, remove clothing immediately and dispose of it in polythene bags
- If any spillage from a battery comes into contact with the eyes, wash repeatedly with eye-wash and seek urgent medical help
- If any spillage from a battery comes into contact with the skin, wash off immediately with water, apply an anti-acid wash, cream or gel to stop burning and then seek urgent medical help
- If you end up with battery acid in your mouth, wash your mouth out with milk. DO NOT swallow the milk. Spit it out. Then seek urgent medical help
- Do not smoke near batteries, and ensure that the area where you are storing the batteries is ventilated
- Prevent arcing or short circuits on battery terminals. Batteries can provide a huge current very quickly. Should you short-circuit a battery with a spanner, the spanner is likely to be red hot within a few seconds and could

easily lead to fire or explosion. Remove any rings, bracelets or watches you may be wearing and keep tools a safe distance away from batteries

Gloves

You need two different sets of gloves for installing your solar array. Chemical gloves for moving batteries and electrical protection gloves for wiring up your solar system.

When choosing suitable chemical gloves for working with batteries, consider the following:

- The gloves need to be quite strong, as lifting and moving batteries is hard on gloves
- A good grip is important
- Buy a glove with a medium or long cuff length, to protect both the hands and forearms
- The gloves should be made of a suitable material to protect against battery acid

The Health and Safety Executive website suggest that 0.4mm-thick neoprene gloves will give suitable protection through a full shift. If you do splash your gloves while working with batteries, make sure you wash them or replace them immediately, to avoid transferring acid to other parts of your body.

Electrically-insulated protection gloves give protection when working with high voltages. These are vitally important when working with high-voltage solar arrays and are recommended for all installations.

Electrically-insulated gloves come with different ratings to provide protection at different voltages:

- Class 00 gloves provide protection for up to 500 volts
- Class 0 gloves provide protection for up to 1,000 volts
- Class 1 gloves provide protection for up to 7,500 volts
- Class 2 gloves provide protection for up to 17,000 volts
- Class 3 gloves provide protection for up to 26,500 volts

For most solar installations, Class 00 or Class 0 gloves are the most appropriate. Remember that the open circuit voltage of a solar array can be more than double the nominal voltage of the solar array: twenty solar panels connected in series may only have a nominal voltage of 240 volts, but the open circuit voltage could be over 500 volts.

Like chemical gloves, choose gloves with a medium or long cuff length to protect both your hands and forearms.

If your electrically-insulated gloves are splashed with battery acid, remove and replace the gloves immediately.

All electrically-insulated gloves should be visually inspected and checked for tears and holes before use. Class 1–3 gloves require full electrical testing every six months.

Electrical safety

I make no apologies for repeating my mantra about electrical safety. Electrical safety is extremely important when installing a solar electric system.

Solar panels generate electricity whenever they are exposed to sunlight. The voltage of a solar panel on an 'open' circuit is significantly higher than the system voltage. A 12-volt solar panel can generate a 22–26 volt current when not connected.

Connect several solar panels in series and the voltage can get to dangerous levels very quickly. A 24-volt solar array can generate 45–55 volts, which can provide a nasty shock in the wrong circumstances, whilst a 48-volt solar array can easily generate voltages of 90–110 volts when not connected. These voltages can be lethal to anyone with a heart condition, or to children, the elderly or pets.

Solar systems produce DC voltage. Unlike AC voltage, if you are electrocuted from a direct current, you will not be able to let go.

Batteries can produce currents measuring thousands of amps. A short circuit will generate huge amounts of heat very quickly and could result in fire or explosion. Remove any rings, bracelets or watches you may be wearing and keep tools away from batteries.

The output from an inverter is AC grid-level voltage and can be lethal. Treat it with the same respect as you would any other grid-level electricity supply.

In many countries, it is law that if you are connecting an inverter into a household electrical system, you must use a qualified electrician to certify your installation.

ASSEMBLING YOUR TOOLKIT

As well as your trusty set of DIY tools, you will need an electrical multi-meter or volt meter to test your installation at different stages. You should use electrically-insulated screwdrivers whilst wiring up the solar array, and a test light circuit tester can be useful.

There are a few sundries that you ought to have as well:

- Cable ties are very useful for holding cables in place. They can keep cable runs tidy and are often good for temporary as well as permanent use
- A water- and dirt-repellent glass polish or wax, for cleaning solar panels
- Petroleum jelly is used on electrical connections on solar panels and batteries to seal them from moisture and to ensure a good connection

PREPARING YOUR SITE

As mentioned in the previous chapter, you may need to consider foundations for ground- or pole-mounting a solar array, or strengthening an existing roof structure if you are installing your solar array on a roof.

If you are installing your batteries in an area where there is no suitable earth (ground), you should install an earthing rod (grounding rod) as close to the batteries as is practical.

TESTING YOUR SOLAR PANELS

Now the fun begins. Start by unpacking your solar PV panels and carry out a visual inspection to make sure they are not damaged in any way.

Chipped or cracked glass can significantly reduce the performance of the solar panels, so they should be replaced if there is any visible damage to the panel. Damage to the frame is not such a problem, so long as the damage will not allow water ingress to the panel and does not stop the solar panel from being securely mounted in position.

Next, check the voltage on the panel using your multi-meter, set to an appropriate DC voltage range.

Solar PV panels generate a much higher voltage on an 'open' circuit (i.e. when the panel is not connected to anything) than they do when connected to a 'closed' circuit. So do not be surprised if your multi-meter records an open voltage of 20–26 volts for a single panel.

INSTALLING THE SOLAR ARRAY

Cleaning the panels

It is a good idea to clean the glass on the front of the panels first, using a water- and dirt-repellent glass polish or wax. These glass polishes ensure that rain and dirt do not stick to the glass, thereby reducing the performance of your solar array. They are available from any DIY store and many supermarkets and car accessories stores.

Assembly and connections

Some roof-mounted solar mounting kits are designed to be fitted to your roof before fitting the solar panels. Others are designed to have the solar panels mounted to the fixing kits before being mounted to the roof.

With a pole-mounted system, you typically erect your pole first and then fit the solar panels once the pole is in position.

A ground-based mounting system is the easiest to install, as there is no heavy lifting.

Typically, you mount and wire the solar panels at the same time. If you are stepping up the voltage of your system by wiring the panels in series, wire up the required number of panels in series first (i.e. sets of two panels for 24 volts, sets of four for 48 volts).

Once you have wired up a set of panels in series, test them using your multi-meter, set to a voltage setting to check that you have the expected voltage (20 volts plus for a 12-volt system, 40 volts plus for a 24-volt system and 80 volts plus for a 48-volt system).

Take care when taking these measurements, as 40 volts and above can give a nasty shock in the wrong circumstances.

Once you have wired each series correctly, make up the parallel connections and then test the entire array using your multi-meter, set to the appropriate voltage setting.

If you have panels of different capacities, treat the different sets of panels as separate arrays. Do not wire panels of different capacities together, either in series or parallel. Instead, connect the arrays together at the controller.

Once you have completed testing, make the array safe so that no one can get an electric shock by accident from the system. To do this, connect the positive and negative cables from the solar array together to short-circuit the array. This will not damage the array and could prevent a nasty shock.

Roof-mounting a solar array

If you are roof-mounting a solar array, you will normally have to fit a rail or mounting to the roof before attaching the solar array.

Once this is in place, it is time to fit the array itself. Make sure you have enough people on hand to be able to lift the array onto the roof without twisting or bending it. Personally, I would always leave this job to professional builders, but the best way seems to be to have two ladders and two people lifting the array up between them, one on each ladder, or using scaffolding.

Final wiring

Once your solar array is in position, route the cable down to where the solar controller is to be installed. For safety purposes, ensure that the cables to the solar array remain shorted whilst you do this.

If you are installing a DC isolation switch and a residual current device (known as a ground fault interrupter in North America), install them between the solar array and the controller.

Once you have the cables in position, un-short the positive and negative cables and check with a meter to ensure you have the expected voltage readings. Then short the cables again until you are ready to install the solar controller.

INSTALLING THE BATTERIES

Pre-installation

Before installing the batteries, you may need to give them a refresher charge before using them for the first time.

You can do this in one of two ways. You can use a battery charger to charge up the batteries, or you can install the system and then leave the solar panels to fully charge up the batteries for a day or so before commissioning the rest of the system.

Put a sticker on each battery with an installation date. This will be useful in years to come for maintenance and troubleshooting.

Positioning the batteries

The batteries need to be positioned so they are upright, cannot fall over and are away from members of the public, children and any sources of ignition.

For insulation and heating purposes, batteries should not be stood directly on a concrete floor. During the winter months, a slab of concrete can get extremely cold and its cooling effects can have detrimental effects on batteries. I prefer to mount batteries on a wooden floor or shelf.

Ventilation

If there is little or no ventilation in the area where the batteries are situated, this must be implemented before the batteries are sited.

As batteries vent hydrogen, which is lighter than air, the gas will rise up. The ventilation should be designed so that the hydrogen is vented out of the battery area as it rises.

Access

It is important that the battery area is easily accessible, not just for installing the batteries (remembering that the batteries themselves are heavy), but also for routinely checking the batteries.

Insulation

As mentioned earlier, if you are installing your batteries in an area that can get very cold or very hot, you should insulate your batteries.

Polystyrene (Styrofoam) sheets or foil-backed bubble-wrap can be used underneath and around the sides of the batteries to keep them insulated. DO NOT INSULATE THE TOP OF THE BATTERIES as this will stop them from venting properly and may cause shorts in the batteries if the insulating material you use is conductive.

Connections

Once the batteries are in place, wire up the interconnection leads between the batteries to form a complete battery bank.

Always use the correct terminals for the batteries you are using and make sure the cables provide a good connection. You should use battery interconnection cables professionally manufactured for the batteries you are using.

Use petroleum jelly around the mountings to seal it from moisture and ensure a good connection.

Next, add an earth (ground) to the negative terminal. If there is no earth already available, install an earthing rod (grounding rod) as close as possible to the batteries.

Now check the outputs at either end of the batteries using a multi-meter to ensure you are getting the correct voltage. A fully-charged battery should be showing a charge of around 13–14 volts per battery.

INSTALLING THE CONTROL EQUIPMENT

The next step is to install the solar controller and the power inverter if you are using one.

Mount these close to the batteries. Ideally they should be mounted within a metre (3 feet), to keep cable runs as short as possible.

Most solar controllers include a small LCD display and a number of buttons to configure the controller. Make sure the solar controller is easily accessible and that you can read the display.

Some solar controllers that work at multiple voltages have a switch to set the voltage you are working at. Others are auto-sensing. Either way, check your documentation to make sure you install the solar controller in accordance with the manufacturer's instructions. If you have to set the voltage manually, make sure you do this now, rather than when you have wired up your system.

Inverters can get very hot in use and adequate ventilation should be provided. They are normally mounted vertically on a wall, to provide natural ventilation. The installation guide that comes with your particular make of inverter will tell you what is required.

Some inverters require an earth (ground) in addition to the earth on the negative terminal on the battery. If this is the case, connect a 2.5mm² green-yellow earth cable from the inverter to your earth rod (ground rod).

If you are installing a DC isolation switch between the solar panels and your control equipment, connect that up first and make sure it is switched off.

Once you have mounted the controller and inverter, connect the negative cables to the battery, taking care that you are connecting the cables to the correct polarity. Then un-short the positive and negative cables from the solar array and connect the negative cable from the solar array to the solar controller, again taking care to ensure the cable is connected to the correct polarity.

Now double-check the wiring. Make sure you have connected the cables to the right places. Double-check that you have connected your negative cable from your solar array to the negative solar input connection on your solar controller. Then double-check that you have connected your negative cable from your battery to the negative battery input on your solar controller and your inverter. Only then should you start wiring up your positive connections.

Start with the battery connection. If you are planning to install a fuse and DC isolation switch into this cable, make sure that your fuse and switch work for both the solar controller and the inverter (if you are using one). Connect the inverter and the solar controller ends first and double-check that you have got your wiring correct, both visually and by checking voltages with a volt meter, before you connect up the battery bank.

Finally, connect up the positive connection from your solar array to the solar controller. At this point, your solar controller should power up and you should start reading charging information from the screen.

Congratulations. You have a working solar power station!

INSTALLING A GRID-TIE SYSTEM

Before starting to install your grid-tie system, you must have already made arrangements with your electricity provider for them to set you up as a renewable energy generator.

Regulations and agreements vary from region to region and from electricity provider to electricity provider, but at the very least they will need to install an export meter to your building to accurately meter how much energy you are providing. They will also ask for an inspection certificate from a qualified electrician, to confirm that the work has been done to an acceptable standard.

Physically, installing a grid-tie system is very similar to installing any other solar energy system, except, of course, you do not have any batteries to work with.

However, you do have to be careful while wiring up the high-voltage solar array. When the solar array is being connected up you can have a voltage build-up of several hundred volts, which can quite easily prove fatal. If building a high-voltage array, cover the solar panels while you are working on them and wear electrically-insulated gloves at all times.

COMMISSIONING THE SYSTEM

Once you have stopped dancing around the garden in excitement, it is time to test what you have done so far and configure your solar controller.

Programming your solar controller

The type of solar controller you have will determine exactly what you need to configure. It may be that you do not need to configure anything at all, but either way you should check the documentation that came with the solar controller to see what you need to do.

Typically, you will need to tell your solar controller what type of batteries you are using. You may also need to tell your solar controller the maximum and minimum voltage levels to show when the batteries are fully charged or fully discharged. You should have this information from your battery supplier, or you can normally download full battery specification sheets from the internet.

Testing your system

You can test your solar controller by checking the positive and negative terminals on the output connectors on the controller using your multi-meter. Switch the multi-meter to DC voltage and ensure you are getting the correct voltage out of the solar controller.

If you have an inverter, plug a simple device such as a table lamp into the AC socket and check that it works.

If your inverter does not work, switch it off and check your connections to the battery. If they are all in order, check again with a different device.

CHARGING UP YOUR BATTERIES

If you have not carried out a refresher charge on your batteries before installing them, switch off your inverter and leave your system for at least 24 hours to give the batteries a good charge.

CONNECTING YOUR DEVICES

Once you have your solar power station up and running and your batteries are fully charged, it is time to connect your devices.

If you are wiring a house using low-voltage equipment, it is worth following the same guidelines as you would for installing grid-voltage circuits. For low-voltage applications, you do not need to have your installation tested by a qualified electrician, but many people choose to do so to make sure there are no mishaps.

The biggest difference between AC wiring and DC wiring is that you do not need to have a separate earth (ground) with DC electrics, as the negative connection is earthed both at the battery and, if you are using one, at the distribution panel.

If you are using 12-volt or 24-volt low-voltage circuits, you can use the same distribution panels, switches and light fittings as you would in a grid-powered home. As already suggested, do not use the same power sockets for low-voltage appliances as you use for grid-powered appliances. If you do, you run the risk that low-voltage appliances could be plugged directly into a high-voltage socket, with disastrous consequences.

IN CONCLUSION

- Once you have done all your preparation, the installation should be straightforward
- Heed the safety warnings and make sure you are prepared with the correct safety clothing and access to chemical clean-up and suitable First Aid in case of acid spills
- Solar arrays are both fragile and expensive. Look after them
- The most likely thing that can go wrong is wiring up something wrongly. Double-check each connection
- Check each stage by measuring the voltage with a multi-meter to make sure you are getting the voltage you expect. If you are not, inspect the wiring and check each connection in turn

TROUBLESHOOTING

Once your solar electric system is in place, it should give you many years of untroubled service. If it does not, you will need to troubleshoot the system to find out what is going wrong and why.

KEEP SAFE

All the safety warnings that go with installation also relate to troubleshooting. Remember that solar arrays will generate electricity almost all the time (except in complete darkness), and batteries do not have an 'off' switch.

COMMON FAULTS

Unless you keep an eye on your solar energy system, problems can often go undetected for months. Only if you have a stand-alone system and the power switches off will you find out that you have a problem.

The faults are typically to be found in one of the following areas:

- Excessive power usage – i.e. you are using more power than you anticipated
- Insufficient power generation – i.e. you are not generating as much power as you expected
- Damaged wiring/ poor connections
- Weak batteries
- Obstructions (shading)
- Faulty earth (ground)
- Inverter faults

Obstructions have been covered in the chapter on surveying your site, starting on page 84. Other faults are covered below.

EXCESSIVE POWER USAGE

This is the most common reason for solar electric systems failing, where the original investigations underestimated the amount of power that was required.

Almost all solar controllers provide basic information on an LCD screen that allows you to see how much power you have generated compared to how much energy you are using, and shows the amount of charge currently stored in the battery bank. Some solar controllers include more detailed information that allows you to check on a daily basis how your power generation and power usage compares.

Using this information, you can check your power drain to see if it is higher than you originally expected.

If you have an inverter in your system, you will also need to measure this information from your inverter. Some inverters have an LCD display and can provide this information, but if your system does not provide this, you can use a plug-in watt meter to measure your power consumption over a period of time.

If your solar controller or your inverter does not provide this information, you can buy a multi-meter with data logging capabilities. These will allow you to measure the current drain from the solar controller and/or your inverter over a period of time (you would typically want to measure this over a period of a day).

Attach the multi-meter across the leads from your batteries to your solar controller and inverter. Log the information for at least 24 hours. This will allow you identify how much power is actually being used.

Some data logging multi-meters will plot a chart showing current drain at different times of the day, which can also help you identify when the drain is highest.

Solutions

If you have identified that you are using more power than you were originally anticipating, you have three choices:

- Reduce your power load
- Increase the size of your solar array
- Add another power source (such as a fuel cell, wind turbine or generator) to top up your solar electric system when necessary

INSUFFICIENT POWER GENERATION

If you have done your homework correctly, you should not have a problem with insufficient power generation when the system is relatively new.

However, over a period of a few years, the solar panels and batteries will degrade in their performance (batteries more so than the solar panels), whilst new obstructions that cut out sunlight may now be causing problems.

You may also be suffering with excessive dirt on the solar panels themselves, which can significantly reduce the amount of energy the solar array can generate. Pigeons and cats are the worst culprits for this!

Your site may have a new obstruction that is blocking sunlight at a certain time of day. For example, a tree that has grown substantially since you carried out the original site survey.

Alternatively, you may have made a mistake with the original site survey and not identified an obstruction. Unfortunately, this is the most common mistake made by inexperienced solar installers. It is also the most expensive problem to fix. This is why carrying out the site survey is so important.

To identify if your system is not generating as much power as originally expected, check the input readings on your solar controller to see how much power has been generated by your solar panels on a daily basis. If your solar controller cannot provide this information, use a multi-meter with data logger to record the amount of energy captured by the solar panels over a three-to-five day period.

Solutions

If you have identified that you are not generating as much power as you should be, start by checking your solar array. Check for damage on the solar array and then give the array a good wash with warm, soapy water and polish using a water- and dirt-repellent glass polish or wax.

Check all the wiring. Make sure that there is no unexplained high resistance in any of the solar panels or on any run of wiring. It could be a faulty connection or a damaged cable that is causing the problems.

Carry out another site survey and ensure there are no obstructions between the solar array and the sun. Double-check that the array itself is in the right position to capture the sun at solar noon. Finally, check that the array is at the optimum angle to collect sunlight.

If you are experiencing these problems only at a certain time of the year, it is worth adjusting the angle of the solar panel to provide the maximum potential power generation during this time, even if this means compromising power output at other times of year.

Check the voltage at the solar array using a multi-meter. Then check again at the solar controller. If there is a significant voltage drop between the two, the resistance in your cable is too high and you are losing significant efficiency as a result. This could be due to an inadequate cable installed in the first place, or damage in the cable. If possible, reduce the length of the cable and test again. Alternatively, replace the cable with a larger and better quality cable.

If none of that works, you have three choices:

- Reduce your power load
- Increase the size of your solar array
- Add another power source (such as a fuel cell, wind turbine or generator) to top up your solar electric system when necessary

DAMAGED WIRING/ POOR CONNECTIONS

If you have damaged wiring or a poor connection, this can have some very strange effects on your system. If you have odd symptoms that do not seem to add up to anything in particular, then wiring problems or poor connections are your most likely culprit.

Examples of some of the symptoms of a loose connection or damaged wiring are:

- A sudden drop in solar energy in very warm or very cold weather. This is often due to a loose connection or damaged wiring in the solar array or between the solar array and the solar controller
- Sudden or intermittent loss of power when you are running high loads. This suggests a loose connection between batteries, or between the batteries and solar controller or inverter
- Sudden or intermittent loss of power on particularly warm days after the solar array has been in the sun for a period of time. This suggests a loose connection somewhere in the array, a damaged panel or high resistance in a cable
- Significantly lower levels of power generation from the solar array suggest a loose wire connection or a short circuit between solar panels within the array
- A significant voltage drop on the cable between the solar array and the solar controller suggests either an inadequate cable or damage to the cable itself

- Likewise, a significant voltage drop on the cable between the solar controller and your low-voltage devices suggests an inadequate cable or damage to the cable itself
- If you find a cable that is very warm to the touch, it suggests the internal resistance in that cable is high. The cable should be replaced immediately

Unfortunately, diagnosing exactly where the fault is can be time-consuming. You will require a multi-meter, a test light and plenty of time.

Your first task is to identify which part of the system is failing. A solar controller that can tell you inputs and outputs is useful here. The information from this will tell you whether your solar array is underperforming or the devices are just not getting the power they need.

Once you know which part of the system to concentrate on, measure the resistance of each cable using the ohm setting on your multi-meter. If the internal resistance is higher than you would expect, replace it. If any cable is excessively hot, replace it. The problem could be caused either by having an inadequate cable in the first place (i.e. too small) or by internal damage to the cable.

Next, check all the connections in the part of the system you are looking at. Make sure the quality of the connections is good. Make sure that all cables are terminated with proper terminators or soldered. Make sure there is no water ingress.

WEAK BATTERY

The symptoms of a weak battery are that either the system does not give you as much power as you need, or you get intermittent power failures when you switch on a device.

In extreme cases, a faulty battery can actually reverse its polarity and pull down the efficiency of the entire bank.

Weak battery problems first show themselves in cold weather and when the batteries are discharged to below 50–60% capacity. In warm weather, or when the batteries are charged up, weak batteries can quite often continue to give good service for many months or years.

If your solar controller shows that you are getting enough power in from your solar array to cope with your loads, then your most likely suspect is a weak battery within your battery bank, or a bad connection between two batteries.

Start with the cheap and easy stuff. Clean all your battery terminals, check your battery interconnection cables, make sure the cable terminators are fitting tightly on the batteries and coat each terminal with a layer of petroleum jelly to ensure good connectivity and protection from water ingress.

Then check the water levels in your batteries (if they are 'wet' batteries). Top up as necessary.

Check to make sure that each battery in your battery bank is showing a similar voltage. If there is a disparity of more than 0.7 volts, it suggests that you may need to balance your batteries.

If, however, you are seeing a disparity on one battery of 2 volts or over, it is likely that you have a failed cell within that battery. You will probably find that this battery is also abnormally hot. Replace that battery immediately.

If your solar controller has the facility to balance batteries, then use this. If not, top up the charge on the weaker batteries, using an appropriate battery charger, until all batteries are reading a similar voltage.

If you are still experiencing problems after carrying out these tests, you will need to run a load test on all your batteries in turn. To do this, make sure all your batteries are fully charged up, disconnect the batteries from each other and use a battery load tester (you can hire these cheaply from tool hire companies). This load tester will identify any weak batteries within your bank.

Changing batteries

If all your batteries are several years old and you believe they are getting to the end of their useful lives, it is probably worth replacing the whole battery bank in one go. Badly worn batteries and new batteries do not necessarily mix well, because of the voltage difference. If you mix new and used, you can easily end up with a bank where some of the batteries never fully charge up.

If you have a bank of part-worn batteries and one battery has failed prematurely, it may be worth finding a second-hand battery of the same make and model as yours. Many battery suppliers can supply you with second-hand batteries. Not only are they much cheaper than new, but because the second-hand battery will also be worn, it will have similar charging and discharging characteristics to your existing bank, which can help it bed down into your system.

If you cannot find a part-worn battery, you can use a new one, but make sure you use the same make and model as the other batteries in your bank. Never mix and match different models of batteries, as they all have slightly different characteristics.

If you add a new battery to a part-worn bank, you may find the life of the new battery is less than you would expect if you replaced all them. Over a few months of use, the performance of the new battery is likely to degrade to similar levels to the other batteries in the bank.

Before changing your battery, make sure that all of your batteries (both new and old) are fully charged.

Put a label on the new battery, noting the date it was changed. This will come in useful in future years when testing and replacing batteries.

Once you have replaced your battery, take your old one to your local scrap merchants. Lead acid batteries have a good scrap value and they can be 100% recycled to make new batteries.

INVERTER ISSUES

The symptoms of inverter issues can include:

- Buzzing or humming sounds from some electronic equipment when powered from the inverter
- Failure of some equipment to run from the inverter
- Regular tripping of circuits
- Sudden loss of power

If you are experiencing buzzing or humming sounds from electronic equipment when powered from the inverter, or if some equipment is not running at all, it suggests that the inverter is not producing a pure AC sine wave. If you have a grid-tie system, the AC pure sine waveform generated by the inverter may not be perfectly coordinated with the waveform from the grid. This would suggest poor quality power from the grid, a grounding issue or a faulty inverter.

If you have a stand-alone system and have purchased a modified sine wave inverter (or quasi-sine wave), it may be that you cannot resolve these issues without replacing the inverter. Some electronic equipment, such as laptop computers and portable televisions, may not work at all using a modified sine wave inverter, whilst other equipment will emit a buzz when run from these inverters.

If you have sudden and unexplained tripping of circuits when running from an inverter, or experience sudden loss of power, there are a number of things to check:

- Does the tripping occur when a heavy-load appliance such as a fridge switches itself on or off?
- Does the tripping occur when the inverter cuts in at the start of the day or when it cuts out at the end of the day?
- Does the tripping occur more often on very warm days or after heavy rain?

Unfortunately, circuit tripping and sudden power loss often only occurs when a combination of events occur, which can make diagnosis time-consuming and difficult to get right.

The most common reasons for circuit tripping or sudden power loss are temperature-related issues. Inverters can generate a huge amount of heat. The hotter they get, the less power they produce. If the inverter is running too hot, a sudden peak demand can be enough to shut the inverter down momentarily. If the inverter runs too hot for too long, it will shut down for a longer period of time to cool.

If this is the case, you are going to have to provide your inverter with more ventilation. If you cannot keep it cool, it may also mean that you require a more powerful inverter to cope with the load.

If the issue occurs during sudden rain or on very hot days, you may also have a grounding problem. Check the inverter with a PAT tester to ensure that you are not getting a ground leakage from the inverter itself. If you are, check all the connections from the DC input of your inverter.

MAINTAINING YOUR SYSTEM

There is very little maintenance to be carried out on a solar electric system. There are some basic checks that should be carried out on a regular basis. Typically these should take no more than a few minutes to carry out.

As required

- Clean the solar array. This actually takes very little effort as unless you live in a very dry and dusty part of the world, the rain will usually do a very good job of washing your solar panels clean on a regular basis
- Telescopic window cleaner kits are available to clean solar arrays mounted on lower sections of a roof. If you can easily access the panels, a dirt- and rain-repellent glass polish can help keep your solar array cleaner for longer
- If you have a thick layer of snow on your solar array, brush it off! A thick blanket of snow very quickly stops your solar array from producing any energy at all

Every month

- If your solar controller or inverter includes a display that shows power input and power output, check the performance of your solar array. Check that it is in line with your expectations for the time of year. It is worth keeping a log of the performance so you can compare it from one year to the next
- If there is an unexplained drop in performance, clean the solar array, visually check the condition of the cables and balance the batteries. If the performance does not improve, follow the troubleshooting guide for further assistance

Every three months

- Check the ventilation in the battery box
- Check the battery area is still weatherproof and there are no leaks from the batteries
- Clean dirt and dust off the top of batteries
- Visually check all the battery connectors. Make sure they are tightly-fitting. Clean and protect them with petroleum jelly where required
- Check the electrolyte level in batteries and top up with distilled water where required

Every six months

- If you have a multi-battery system and your solar controller has the facilities to do so, balance the battery bank
- Once the batteries are balanced, use a volt meter or multi-meter to check the voltage on each individual battery. Ensure the voltages are within 0.7 volts of each other
- If one or more battery has a big difference in voltage, follow the instructions on weak batteries in the troubleshooting section of this book

Every year

- If you have a battery bank with more than two batteries, swap the order of the batteries in your bank. Place the batteries that were in the middle of your bank at either end and the batteries that were at either end of your bank in the middle. Then balance them
- This will ensure that all the batteries get even wear throughout their lifetime and thereby increase the overall lifespan of the batteries

At the start of each winter

- Check the insulation around the batteries
- Check that the area around the batteries is free of rodents. Mice and rats like to keep warm, and insulation around batteries is a tempting target. If they have found your batteries, they are likely to gnaw the cabling as well
- Clean the solar array to ensure you get the best possible performance at the worst time of the year

INTERNET SUPPORT

A free website supports this book. It provides up-to-the-minute information and online solar energy calculators to help simplify the cost analysis and design of your solar electric system.

To visit this site, go to the following address:

www.SolarElectricityHandbook.com

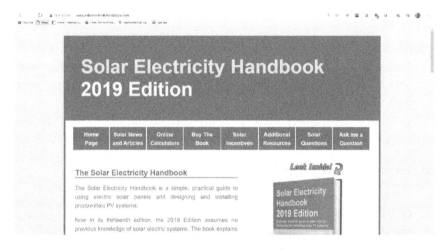

TOOLS AVAILABLE ON THE WEBSITE

Online project analysis

The online project analysis tool on this site takes away a lot of the calculations that are involved with designing a new solar electric system, including estimating the size and type of solar panel, the size and type of battery, the thickness of low-voltage cable required and providing cost and timescale estimates.

To use the online project analysis, you will need to have completed your power analysis (see the chapter on Project Scoping), and ideally completed your whole project scope. The solar calculator will factor in the system inefficiencies and produce a thirteen-page analysis for your project.

Monthly insolation figures

Monthly solar insolation figures for every country in the world are included on the website. These can be accessed by selecting your country and the name of your nearest town or city from a list. Every country in the world is included.

The solar insolation figures used are monthly averages based on three-hourly samples taken over a 22-year period.

Solar angle calculator

The solar angle calculator shows the optimum angle for your solar panel on a month-by-month basis, and shows where the sun will rise and set at different times of the year.

Solar resources

A directory of solar suppliers is included on the website, along with links for finding out the very latest about grant schemes and selling your electricity back to the electricity companies.

Questions and answers

The site includes an extensive list of questions posted on the site by other site visitors, along with my answers. These questions and answers cover almost every conceivable area of solar design and installation and are worth a browse. If you have a question of your own, post it on the site.

Author online!

If you have noticed a mistake in the book, or feel a topic has not been covered in enough detail, I would welcome your feedback.

This handbook is updated on a yearly basis and any suggestions you have for the next edition would be gratefully received.

The website also includes an 'ask me a question' facility, so you can get in touch with any other questions you may have, or simply browse through the questions and answers in the Frequently Asked Questions section.

Solar articles

New articles about solar power are regularly added to our articles section.

APPENDIX A – TYPICAL POWER REQUIREMENTS

When creating your power analysis, you need to establish the power requirements for your system. The best way is to measure the actual power consumption using a watt meter.

Finding a ballpark figure for similar devices is the least accurate way of finding out your true power requirements. However, for an initial project analysis it can be a useful way of getting some information quickly:

HOUSEHOLD AND OFFICE

Device	Watts	Device	Watts
Air conditioning	2500	– Server – large	220
Air cooling	700	– Server – small	120
Cell phone charger	10	Deep fat fryer	1450
Central heating pump	800	Dishwasher	1200
Central heating controller	20	Electric blanket – double	100
Clothes dryer	2750	Electric blanket – single	50
Coffee maker – espresso	1200	Electric cooker	4000
Coffee percolator	600	Electric toothbrush	1
Computer systems:		Fan – ceiling	80
– Broadband modem	25	Fan – desk	60
– Broadband and wireless	50	Fish tank	5
– Desktop PC	80	Food mixer	130
– Document scanner	40	Fridge – 12 cu. ft.	280
– Laptop	30	Fridge – caravan fridge	110
– 17" flat screen	18	Fridge – solar energy saving	5
–19" flat screen	22		
–22" flat screen	24	Fridge-freezer – 16 cu. ft.	350
– Netbook	9	Fridge-freezer – 20 cu. ft.	420
– Network hub – large	20	Hair dryer	1000
– Network hub – small	8	Heater – fan	2000
– Inkjet printer	150	Heater – halogen spot heater	1000
– Laser printer	250		

Device	Watts	Device	Watts
Heater – oil filled radiator	1000	Power shower	2400
Heater – underfloor (per m²)	150	Radio	10
		Sewing machine	75
Iron	1000	Shaver	6
Iron – steam	1500	Slow cooker	150
Iron – travel	600	Television:	
Kettle	2000	– LCD 15"	15
Kettle – travel	700	– LCD 20"	19
Lightbulb – LED	3	– LCD 24"	23
Lightbulb – CFL	11	– LCD 32"	32
Lightbulb – fluorescent	60	– DVD player	10
Lightbulb – halogen	50	– Set top box	25
Lightbulb – incandescent	60	– Video games console	45
Microwave oven – large	1400	Toaster	1200
Microwave oven – small	900	Upright freezer	250
Music system – large	250	Vacuum cleaner	700
Music system – small	80	Washing machine	550
Photocopier	1000	Water heater – immersion	3000

GARDEN AND DIY

Device	Watts	Device	Watts
Concrete mixer	1400	– LED	10
Drill:		Hedge trimmer	500
– Bench drill	1500	Lathe – small	650
– Hammer drill	1150	Lathe – large	900
– Handheld drill	700	Lawn mower:	
– Cordless drill charger	100	– Cylinder mower – small	400
Electric bike charger	100	– Cylinder mower – large	700
Flood light:		– Hover mower – small	900
– Halogen – large	500	– Hover mower – large	1400
– Halogen – small	150	Lawn raker	400
– Fluorescent	36	Pond:	

Device	Watts	Device	Watts
– Small filter	20	Jigsaw	550
– Large filter	80	Angle grinder – small	1050
– Small fountain pump	50	Angle grinder – large	2000
– Large fountain pump	200	LED Shed light	5
Rotavator	750	Strimmer – small	250
Chainsaw	1150	Strimmer – large	500

CARAVANS, BOATS AND RECREATIONAL VEHICLES

Device	Watts	Device	Watts
Air cooling	400	– Electric/gas fridge	60
Air heating	750	– Low energy solar fridge	10
Coffee percolator	400	Kettle	700
Fridge:		Fluorescent light	10
– Cool box – small	50	Halogen lighting	10
– Cool box – large	120	LED lighting	3

APPENDIX B – LIVING OFF-GRID

Living off-grid is an aspiration for many people. You may want to 'grow your own' electricity and not be reliant on electricity companies. You may live in the middle of nowhere and be unable to get an outside electricity supply. Whatever your motive, there are many attractions for using solar power to create complete self-sufficiency.

Do not confuse living off-grid with a grid-tie installation and achieving a balance where energy exported to the grid minus energy imported from the grid equals a zero overall import of electricity. A genuinely off-grid system means you use the electricity you generate every time you switch on a light bulb or turn on the TV. If you do not have enough electricity, nothing happens.

Be under no illusions: this is going to be a big lifestyle project and for most people it will involve making some significant compromises on power usage to make living off-grid a reality. In this book, I have been using the example of a holiday home. The difference between a holiday home and a main home is significant. If you are planning to live off-grid all the time, you may not be so willing to give up some of the creature comforts that this entails. Compromises that you may be prepared to accept for a few days or weeks may not be so desirable for a home you are living in for fifty-two weeks a year.

Remember that a solar electric system is a long-term investment, but will require long-term compromises as well. You will not have limitless electricity available when you have a solar electric system, and this can mean limiting your choices later on. If you have children at home, consider their needs as well. They will increase as they become teenagers and they may not be so happy about making the same compromises that you are.

You need to be able to provide enough power to live through the winter as well as the summer. You will probably use more electricity during the winter than the summer: more lighting and more time spent inside the house mean higher power requirements.

Most off-grid installations involve a variety of power sources, such as a solar electric system, a wind turbine and possibly a hydro-generation system if you have a fast-flowing stream with a steep enough drop. Of these technologies, only hydro on a suitable stream has the ability to generate electricity 24 hours a day, seven days a week. In addition to using solar, wind and hydro for electrical generation, a solar water system will help heat up water and a ground source heat pump may be used to help heat the home.

When installing these systems in a home, it is important to have a *failover* system in place. A failover is simply a power backup so that if the power generation is insufficient to cope with your needs, a backup system cuts in. Diesel generators are often used for this purpose. Some of the more expensive solar controllers have the facility to work with a diesel generator, automatically starting up the generator to charge up your batteries if the battery bank runs too low on power. Advanced solar controllers with this facility can link this in with a timer to make sure the generator does not start running at night when the noise may be inappropriate.

A SOLAR ELECTRIC SYSTEM IN CONJUNCTION WITH GRID ELECTRICITY

Traditionally, it has rarely made economic sense to install a solar electric system for this purpose. This has changed over the past three years, with the availability of financial assistance in many parts of the world.

If you are considering installing a system purely on environmental grounds, make sure that what you are installing actually does make a difference to the environment. If you are planning to sell back electricity to the utility grids during the day, then unless peak demand for electricity in your area coincides with the times your solar system is generating electricity, you are actually unlikely to be making any real difference whatsoever.

A solar energy system in the southern states of America can make a difference to the environment, as peak demand for electricity tends to be when the sun is shining and everyone is running air conditioning units. A grid-tie solar energy system in the United Kingdom is unlikely to make a real difference to the environment unless you are using the electricity yourself or you live in an industrial area where there is high demand for electricity during the day.

If you are in the United Kingdom or Canada and are installing a solar energy system for the primary motive of reducing your carbon impact, a grid fallback system is the most environmentally friendly solution. In this scenario, you do not export energy back to the grid, but store it and use it yourself. When the batteries have run down, your power supply switches back to the grid. There is more information on this configuration later in this chapter.

There may be other factors that make solar energy useful. For example, ensuring an electrical supply in an area with frequent power cuts, using the solar system in conjunction with an electric car, or for environmental reasons where the environmental benefits of the system have been properly assessed.

One of the benefits of building a system to work in conjunction with a conventional power supply is that you can take it step by step, implementing a small system and growing it when finances allow.

As outlined in chapter three, there are three ways to build a solar electric system in conjunction with the grid: a grid-tie system, a grid interactive system and grid fallback. You can choose to link your solar array into the grid as a grid-tied system if you wish, so that you supply electricity to the grid when your solar array is generating the majority of its electricity and you use the grid as your battery. It is worth noting that if there is a power cut in your area, your solar electric system will be switched off as a safety precaution, which means you will not be able to use the power from your solar electric system to run your home, should there be a power cut.

Alternatively, you can design a stand-alone solar electric system to run some of your circuits in your house, either at grid-level AC voltage or on a DC low-voltage system. Lighting is a popular circuit to choose, as it is a relatively low demand circuit to start with.

GRID FALLBACK AND GRID FAILOVER SYSTEMS

As a third alternative, you can wire your solar electric system to run some or all of the circuits in your house, but use an AC relay to switch between your solar electric system when power is available, and electricity from the grid when your battery levels drop too low. In other words, you are using the grid as a power backup, should your solar electric system not provide enough power. This setup is known as a grid fallback system.

Grid fallback and grid failover are both often overlooked as a configuration for solar power. Both these systems provide AC power to a building alongside the normal electricity supply and provide the benefit of continued power availability in the case of a power cut. For smaller systems, a solar electric emergency power system can be cost-competitive with installing an emergency power generator and uninterruptable power supplies. A solar electric emergency power system also has the benefit of providing power all of the time, thereby reducing ongoing electricity bills as well as providing power backup.

The difference between a grid fallback system and a grid failover system is in the configuration of the system. A grid fallback system provides solar power for as much of the time as possible, only switching back to the grid when the batteries are flat. A grid failover system cuts in when there is a power cut.

Most backup power systems provide limited power to help tide premises over a short-term power cut of 24 hours or less. Typically, a backup power system would provide lighting, enough electricity to run a heating system and enough electricity for a few essential devices.

As with all other solar projects, you must start with a project scope. An example scope for a backup power project in a small business could be to provide electricity for lighting and for four PCs and to run the gas central heating for a maximum of one day in the event of a power failure. If your premises have a number of appliances that have a high-energy use, such as open fridges and freezer units, for example, it is probably not cost-effective to use solar power for a backup power source.

Installing any backup power system will require a certain amount of rewiring. Typically, you will install a secondary distribution panel (also known as a consumer unit) containing the essential circuits, and connect this after your main distribution panel. You then install an AC relay or an Automatic Transfer Switch between your main distribution panel and the secondary distribution panel, allowing you to switch between your main power source and your backup source:

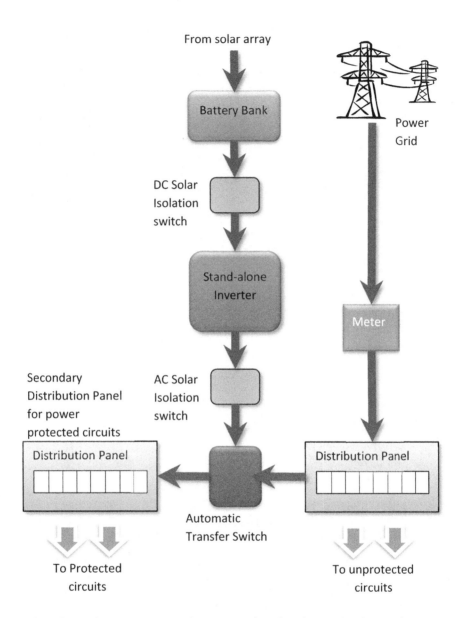

In this above diagram, a second consumer box has been wired into the electrical system, with power feeds from both the main consumer box and an inverter connected to a solar system.

Switching between the two power feeds is an automatic transfer box. If you are configuring this system to be a grid fallback system, this transfer box is configured to take power from the solar system when it is available, but then switches back to grid-sourced electricity if the batteries on the solar system have run down.

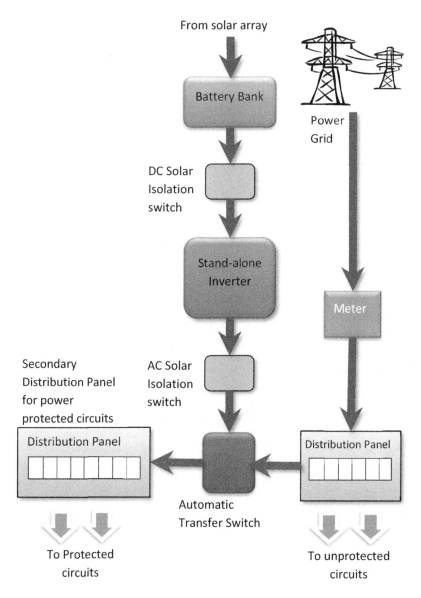

This provides a backup for critical power when the normal electricity supply is not available, but also uses the power from the solar system to run your devices when this is available.

If you are configuring this system to be a grid failover system, the transfer box is configured to take power from the normal electricity supply when it is available, but then switches to power from the solar system if it is not.

One issue with this system is that when the transfer box switches between one power source and the other, there may be a very short loss of power of around $1/20$ of a second. This will cause lights to flicker momentarily, and in some cases may reset electronic equipment such as computers, TVs and DVD players.

Many modern transfer boxes transfer power so quickly that this is not a problem. However, if you do experience this problem it can be resolved by installing a small uninterruptable power supply (UPS) on any equipment affected in this way.

You can buy fully built-up automatic transfer boxes, or you can build your own relatively easily and cheaply using a high-voltage AC Double-Pole/Double-Throw (DPDT) Power Relay, wired so that when the inverter is providing power, the relay takes power from the solar system, and when the inverter switches off, the relay switches the power supply back to the normal electricity supply.

A FINAL WORD

It is almost twenty-four years since I built my first solar array. Since then, I have not always been working with solar technology full time, but it is a technology I come back to time and time again, and my enthusiasm for solar is as strong now as it has always been.

Solar power is an excellent and practical resource. It can be harnessed relatively easily and effectively. It inspires people and is making a huge contribution to transforming lives around the world.

From an enthusiast's perspective, designing and building a solar electric system from scratch is interesting, educational and fun. If you are tempted to have a go, start with something small like a shed light, and feel free to experiment with different ideas.

Solar is making the seemingly impossible possible for the first time. In 2011 and 2012, I was involved in a project to provide solar power to remote villages in Africa, providing a reliable source of electricity for lighting, communications and the ability to store medication safely at chilled temperatures.

In 2014, my solar project was working on a solar electric delivery van for use in big cities. The prototype vehicle was so efficient it provided a solar-only range of around 15 miles (25km) in dull overcast winter conditions in London, or up to over 100 miles of solar range in California on a summer day. More recently, I led a research and development team developing solar battery systems and the next generation of solar panels.

Today, I am working on battery storage systems as a way of rapidly charging electric vehicles. Again, solar is an integral part of our new system.

If I have inspired you to do something with solar energy yourself, then it has served its purpose. I wish you the very best for your project.

All the best

Michael Boxwell
January, 2021

Made in the USA
Las Vegas, NV
12 December 2021